Capacity Building in Fisheries and Aquaculture

NIPA® GENX ELECTRONIC RESOURCES & SOLUTIONS P. LTD.
New Delhi-110 034

About the Editors

Dr. B. P. Mohanty, Asst. DG (Inland Fisheries) ICAR since March 2020. MVSc (Biochemistry) from IVRI, Izatnagar and PhD from JNU, New Delhi. Started his career as Asst. Professor (Biochemistry) in OUAT, Bhubaneswar in Mar 1990. Joined ARS, ICAR in 1994. He Served as Scientist, Senior Scientist, Principal Scientist and Head of Division at CIFRI, Barrackpore. He was the Conferred Biotechnology Overseas Associateship Award (BTOA) by DBT, Government of India for advanced training in Proteomics. He Published more than 100 research papers in journals of repute and has three Books, 10 Research Bulletins, two Open-access Database (NutriFishIN & FISHPROT) and one Mobile App (NutrifishApp) to his credit. He is a Life Member and Fellow of several prestigious scientific societies and academies. He is Fellow of National Academy of Agricultural Sciences (FNAAS), Zoological Society Kolkata (FZS), Inland Fisheries Society of India (FIFSI), Zoological Society of India (FZSI); Hon. He is Fellow of International Society for Ecological Communications (FISEC) and Life Member NASI Allahabad. Received Gold Medal from NAARM, Hyderabad; Zoological Society of India and OP Odum Gold Medal from FISEC & Lifetime Achievement Award in Biochemistry. Member Editorial Advisory Board, CAB International (CABI), UK. Member Board of Management, OUAT Bhubaneswar, Odisha & TNJFU Nagapattinum, Tamil Nadu. Dr. Mohanty is currently the President of Society of Veterinary Biochemists and Biotechnologists of India (SVBBI).

Dr. Archana Sinha, Head and Principal Scientist, Aquatic Environmental Biotechnology Division, ICAR-Central Inland Fisheries Research Institute (ICAR-CIFRI), Barrackpore, Kolkata has > 30 years experience in Fisheries and aquaculture research, education and extension. She was a course teacher for M. F. Sc., Ph.D. and Diploma course on Ornamental fish. She developed course module on ornamental fish breeding and culture for ICAR-CIFE (Deemed University), Mumbai; Skill Development Programme of ASCI, New Delhi. She is a recognized expert on Ornamental fish breeding and culture . She organized more than 100 short term training programmes on different activities of fisheries and aquaculture for unemployed, women, farmers and entrepreneurs. She has published more than 100 research papers, 6 books, a series of technical and extension bulletins. She has been conferred with many awards like Best Scientist award by Heeralal Chaudhuri Fisheries Foundation, Mumbai; NATCON Environmental Conservation Award by Nature Conservator, Muzaffarnagar; Gold Medal by Zoological Society of India.

Dr. Shahaji Phand, Deputy Director and Centre Head of Centre for Extension in Agri-allied Sector and also Principal Coordinator, Agri-Clinics and Agribusiness Scheme (AC&ABC), National Institute of Agricultural Extension Management (MANAGE), Hyderabad. He is veterinary graduate and completed Bachelor of Veterinary Science and Animal Husbandry (B. V. Sc. A.H) degree in 2002 from Maharashtra Animal and Fishery Science University, Nagpur Maharashtra. He completed his Master and Doctoral Degree in Veterinary Extension Education from Indian Veterinary Research Institute, Izatnagar, Bareilly, Uttar Pradesh in 2005 and 2008 with Junior and Senior Research Fellowship respectively. After completion of doctoral degree, he joined as Assistant Professor in Nanaji Deshmukh Veterinary Science University (NDVSU) in June 2008, promoted as Associate Professor in 2012 in same university. After serving 8 years as faculty in university, he joined as Assistant Director in MANAGE, Hyderabad in 2016 and promoted to post of Deputy Director in 2018. He has developed 02 technologies in the form of computer software's namely Animal Health Information System (AHIS) and Health Information System for Dairy Animal (HIS) as part his doctoral research, which is patented and commercialized. For this contribution he has been conferred with Dr. C.M. Singh Award for Best Ph. D Research at national level. He has published more than 30 research and review papers in national and international journal and edited of more than 27 books. He has been editor, reviewer and member of editorial board of national and international Journals. He has organized and coordinated more than 300 training programs for senior and middle level officers of State Department of Agriculture and Allied sector, Scientist, SAU faculties, Students etc. which benefit reached more than 4000 stakeholders. He has organized and participated national level conferences, seminar, workshop and contributed for policy formulation. As a Principal Coordinator, he is implementing Agri-Clinics and Agribusiness Scheme (AC&ABC) with the help of 126 Nodal Training Institute (NTIs) across the country, which have trained more than 7500 agricultural graduates and diploma holder, among which more than 3000 trained candidates have established their agri-venture in last two years and serving farming community.

Dr. Sushrirekha Das presently working as MANAGE Fellow in extension in Agri-allied Centre, MANAGE, Hyderabad, India in the Fishery sector. She has completed her Graduation in Zoology as a topper from Swami Vivekanda Autonomous College, Jagatsinghpur, completed her M.Sc in Zoology with Specialization in Fisheries Science and did her Ph.D. in Zoology with research specialization in Fisheries Science. Her core research is in probiotics from Sriram Chandrabhanj

Deo University, Mayurbhanj, Baripada, Odisha. She has put 3 Years of service in teaching as Zoology lecturer. She is having 5 years as Fishery extension officer. She is having more than 8 years of experience in the field of education, research and extension in aquaculture sector. She has submitted 15 numbers of genes to the NCBI Gen Bank. She has published patent in Probiotic rich fermented Amla (Embilica officinalis) fruits and a method of preparation thereof. She has developed 4 probiotics aquaculture products. She has guided 4 nos. of Master degree and graduate students, and handled one Research Projects in SERB, India. She has organized and coordinated more than 70 Workshops, Training Programmes on aquaculture and animal husbandry sector by MANAGE. She has trained more than 500 fish and Shrimp farmers of Odisha and other states through various funding agencies such as NFDB and MANAGE. She has coordinated 115 nos. of SCSP training program in the Fy- 2021-22, there are 4805 fishermen, fisher women, aquapreneurs, students and research scholars in PAN India. She has published more than 20 Research papers of National and International Repute and more than 8 popular articles. She has published 22 Chapters in different books. She has contributed as editor in 26 e-books in agricultural extension. She has received many prestigious Awards such as Young Scientist Award, Best poster/oral presentation Award, Best Performer Award and she has awarded with best service provider award in fishery extension service and Rastriya Samaj Seva Ratna Award etc.

Dr. Yasmeen Basade is working as Principal Scientist (Inland Fisheries) in the Fisheries Science Division, Indian Council of Agricultural Research (ICAR HQs), New Delhi, India. She was involved in the World Bank funded, National Agricultural Innovation Project (NAIP) at Project Implementation Unit, ICAR HQs. Dr. Basade has also served in the ICAR - Directorate of Coldwater Fisheries Research (DCFR), Bhimtal, Uttarakhand. Dr. Basade earned her PhD from ICAR - Central Institute of Fisheries Education (CIFE), Mumbai. She has extensive experience in the domain of coldwater fisheries focusing on fish nutrition, aquaculture and resource assessment, coldwater fish breeding and seed production, cage culture in upland lakes. She has operated numerous research projects as Principal Investigator and has held various leadership roles in capacity as Officer-in-Charge, Nodal Officer, Member Secretary, Organizing Secretary, Course Coordinator. She has a good number of publications. She has also served as referee for peer-reviewed journals in field of fisheries science.

Capacity Building in Fisheries and Aquaculture

Bimal Prasanna Mohanty
Archana Sinha
Shahaji Phand
Sushrirekha Das
Yasmeen Basade

NIPA® GENX ELECTRONIC RESOURCES & SOLUTIONS P. LTD.
New Delhi-110 034

NIPA® GENX ELECTRONIC RESOURCES & SOLUTIONS P. LTD.

101,103, Vikas Surya Plaza, CU Block
L.S.C.Market, Pitam Pura, New Delhi-110 034
Ph. +91 11 27341616, 27341717, 27341718
E-mail: newindiapublishingagency@gmail.com
www: www.nipabooks.com

For customer assistance, please contact
Phone: + 91-11-27 34 17 17
Fax: + 91-11-27 34 16 16
E-Mail: feedbacks@nipabooks.com

Print ISBN: 978-93-58875-61-4

ebook ISBN: 978-93-58878-16-5

Composed and Designed by NIPA®.

Dedicated to

Dr. Trilochan Mohapatra

Chairperson

Protection of Plant Varieties and

Farmers' Rights Authority (PPVFRA)

Government of India, New Delhi

&

Former Secretary DARE &

Director General ICAR

Ministry of Agriculture & Farmers Welfare

Government of India, New Delhi

डॉ त्रिलोचन महापात्र
अध्यक्ष
पौधा किस्म और कृषक अधिकार संरक्षण प्रधिकरण
(संसद क अधिनियम द्वारा निर्मित संविधिक निकाय)
कृषि एवं किसान कल्याण मंत्रालय
भारत सरकार

Dr. Trilochan Mohapatra
Chairperson
Protection of Plant Varieties and Farmers' Rights Authority
(A Statutory body created by an Act of Parliament)
Ministry of Agriculture and Farmers Welfare
Government of India

Foreword

Fisheries, being one of the fastest growing segments of agriculture of our country, is reckoned as a sunrise sector of the Indian economy. Fisheries and aquaculture has emerged as a leading source of food, nutrition, employment and revenue generation. It ensures availability of good quality animal protein at an affordable price, particularly to the economically weaker sections of the society. The sector engages millions of fishers and fish farmers at the primary level including a large number of economically disadvantaged and vulnerable population along the value chain. With the available wide-ranging diversified fisheries resources across the country in the form of marine and inland water bodies, fisheries and aquaculture is emerging consequently as a promising area with immense growth potential.

Utilizing the resources judiciously for sustainable production from fisheries and aquaculture needs constant innovation and diffusion of technologies to the end users. The sector, however laces various challenges due to climate change and declining land and water resources. Besides, insufficient capacity among the Ushers and fish farmers persists as one of the critical challenges to be addressed. Lack of adequate capacity building deprives fishers and fish farmers of the benefits of promising research findings to enhance fish production and productivity, improve methods of storage and transportation to increase shelf life and reduce post¬harvest losses, develop value-added products and byproducts to get better returns, and conserve resources to get sustainable returns.

Effective capacity building is critical for sustainable development of fisheries and particularly important for livelihood, food and nutritional security of fishers and fish farmers. A successful capacity building programme demands a careful functional design and a process involving all the stakeholders. 1 his document is a compendium of a few capacity building activities sponsored by national and international organizations such as Indian Council of Agricultural Research (ICAR), National Fisheries Development Board (NFDB), National Institute of Agricultural Extension Management (MANAGE). World Fish and United States Agency for International Development, (USAID) with the aim to reach all those committed to undertaking capacity building for sustainable fisheries and rural development. The programmes are conceived to strengthen and empower fishers and fish farmers to attain sustainable livelihood for themselves and become agents of change for their community. Emphasis is also placed on generating mass awareness on nutritional benefits of fish, reducing post-harvest losses, requirements of hygiene, packaging and quality certifications, sensitizing farmers about FPOs to help them improve access to technology and resources as well as overcome the constraints they face as individual.

This document offers concepts, tools and best practices that can be followed in Older to help its users obtain a better understanding of how capacity building programmes can be developed to perform effectively. 1 lope this publication will be a useful source of motivation for the academicians and extension workers, and serve us a key resource for planners, policy makers and financial institutions addressing food, nutrition and livelihood related challenges.

I compliment the authors for their efforts and valuable work.

Date 1st April. 2024

Trilochan Mohapatra

एन ए एस सी काम्पैलैक्स, डी पी एस मार्ग, नई दिल्ली - 110012
NASC Complex, DPS Marg, New Delhi-110012
Tel: 011-25848127, 25841696; Mob : +91-7840000542; Fax: 011-25840478
Website: www.plantauthority.gov.in
E-mail: chairperson-ppvfra@nic.in

डॉ पी चन्द्रा शेखरा
महानिदेशक

Dr. P. Chandra Shekara
Director General

Message

It is my pleasure to introduce book entitled "Capacity Building in Fisheries and Aquaculture" for sustainable development of sector. The present book navigating towards harnessing the potential of fishery sector through enhancing knowledge and skill. The fisheries and aquaculture play a crucial role in providing food security, livelihoods, and economic growth for millions of people worldwide. However, the sustainability of the sector is under increasing pressure due to factors such as; overexploitation, habitat degradation climate change etc. To address these challenges, there is need of multifaceted approaches, along with technological innovation, bringing the awareness, improvement in knowledge and skills are also one of the important aspect.

This book serves as a comprehensive guide for researchers, extension officials and stakeholders seeking to enhance capacity in the fisheries and aquaculture sector. Through a combination of theoretical frameworks, case studies, and practical tools, it offers actionable strategies for building the knowledge, skills, and institutional frameworks necessary for sustainable development.

Furthermore, "Capacity Building in Fisheries and Aquaculture" underscores the importance of collaboration and knowledge exchange among stakeholders at all levels. By fostering partnerships between government agencies, NGOs, academia, and local communities, we can leverage collective expertise and resources to address the complex challenges facing in oceans and coastal areas.

I extend my heartfelt thanks to the authors, contributors, and editors who have worked tirelessly to bring this valuable resource to fruition. May it inspire and empower all who seek to chart a course towards a more sustainable and equitable future for the fishery sector.

Dr. P. Chandra Shekara
Director General, MANAGE

राष्ट्रीय कृषि विस्तार प्रबंध संस्थान (मैनेज)
(कृषि एवं किसान कल्याण मंत्रालय, भारत सरकार का एक स्वायत्त संगठन, राजेन्द्रनगर, हैदराबाद - 500 030, तेलंगाना, भारत.)
NATIONAL INSTITUTE OF AGRICULTURAL EXTENSION MANAGEMENT (MANAGE)
(An Autonomous Organization of Ministry of Agriculture and Farmers Welfare, Government of India)
Rajendranagar, Hyderabad - 500 030, Telangana State, INDIA

Acknowledgements

- Dr. Himanshu Pathak, Secretary DARE and Director General ICAR for the encouragement and support
- Dr. P. Chandra Shekara, Director General MANAGE for the encouragement and support
- Dr. J. K. Jena, Deputy Director General (Fisheries Science) for the encouragement
- The NIPA Team led by Mr. Sumit Jain for making it possible
- Mr. Divyanshu Mohanty, B. Tech - Chemical Engineering (Final Year), NIT Rourkela for the editorial help
- Our families for their time and emotional support

Preface

India is an agrarian economy; about 55% of the total work force isengaged in agriculture and allied sector activities (Census 2011). The people belonging to the farming community, whether in agriculture, horticulture, aquaculture, fisheries, etc. are the main actors in improving agricultural production and productivity, and therefore, empowering them with skill, knowledge, training, and mass awareness programmes (collectively called capacity building, CB) would ensure food and nutritional security by alleviating hunger and malnutrition. Such knowledge-empowered farmers, in combination with the right government policies, would lead towards achieving the Sustainable Development Goals (SDGs), especially Goal #2: Zero Hunger by 2030.

Keeping such facts in view, the Indian Council of Agricultural Research (ICAR) and the Department of Agricultural Research and Education (DARE) under the Ministry of Agriculture and Farmers' Welfare Government of India, follow the policy of "Farmers First" and lay great emphasis on capacity building through their institutional mechanisms.

In this document, a modest attempt has been made to compile and highlight a few such CB programmes sponsored by national and international agencies like ICAR, NFDB (National Fisheries Development Board under the Ministry of Fisheries, Animal Husbandry, and Dairying), MANAGE (National Institute of Agricultural Extension Management), World Fish, USAID (United States Agency for International Development), and other institutional mechanisms.

Jai Kisan, Jai Vigyan **Editors**

Contents

1

Mass Awareness Program on Benefits of Fish Consumption

Bimal Prasanna Mohanty

Indian Council of Agricultural Research (ICAR), New Delhi, Delhi, India

Health benefits of fish consumption

Fisheries and aquaculture play an important role in providing food and nutritional security. Fish is a rich source of quality animal proteins, omega-3: poly-unsaturated fatty acids (PUFAs), minerals, and vitamins. Fish is a superfood that is rich in all nutrients except carbohydrates, and its combination with rice or wheat makes it a complete food, forming a balanced diet with the right combination of macronutrients (quality proteins, fats, and carbohydrates) and micronutrients (minerals and vitamins), which are essential for good health.

Aquaculture is one of the fastest-growing food production sectors. India is bestowed with all types of aquatic habitats; freshwater, including the cold waters in the highlands, brackish waters, and the marine system. Being a tropical country, India is a biodiversity hotspot. It is rich in a wide variety of fish fauna, and wherever there is water, there are fish. Thus, fish is widely available in India, Thus, India is the third largest fish producing country and ranks second in aquaculture production. In comparison to other sources of animal proteins, fish is more chicken, mutton, beef, pork, etc., as affordable is available in wide varieties in the markets. Therefore, the common man can afford to consume fish regularly and provide his family with a balanced and healthy diet.

As a diet, fish is important for all age groups of humans. Fish oil, especially the omega-3 PUFAs, EPA, and DHA, keeps away several preventable human ailments (Table 1).

Table 1: Role of fish oils (omega-3 polyunsaturated fatty acids) in preventing human ailments

Population	Preventable diseases
Adult	Coronary heart disease (CHD)
Geriatric (Elderly)	Dementia
	Age-related macular degeneration (AMD)
	Mood disorders
Paediatric	Attention-deficit hyperactivity disorder (ADHD)
	Childhood asthma

As a rich source of quality animal proteins, fish has a good balance of essential and non-essential amino acids. Proteins play several important physiological roles as signalling molecules, precursors of hormones, transport proteins, and immune molecules, including immunoglobulins(Table 2).

Table 2: Multiple functions of proteins with examples

Proteins	1. Structure and mechanics (Actin, Myosin, Troponin)
	2. Transport function (Albumin)
	3. Enzymes (Trypsin, Pepsin, DNA Pol.)
	4. Channels and pumps (Band-3 protein, Na^+ K^+-ATPase).
	5. Hormones (Insulin, Glucagon)
	6. Signaling molecules (Adenylate cyclase), cytokines)
	7. Acid-base balance (Plasma proteins)
	8. Fluid balance (Albumin, vasopressin)
	9. Antibodies (Immunoglobuli, Cytokines)

Fish, especially small indigenous fish (SIFs), are micronutrient-dense; SIFs contain as many as 5% minerals, compared to 0.5% in some large fish. These minerals play an important role in enzyme catalysis, thus playing a modulatory role in several metabolic processes. Some important minerals, with their roles as enzyme cofactors, are listed below (Table 3).

Table 3: Some important minerals that serve as cofactor(s) for enzymes

Inorganic elements	Enzymes
Fe^{2+} or Fe^{3+}	Catalase, Cytochrome-c reductase, Nitrogenase, Hydrogenase, L-arginine hydroxylase, Superoxide dismutase Tryptophan dehydrogenase, Calcium/calmodulin-dependent protein kinase, Inositol-trisphosphate 3-kinase
Cu^{2+}	Cytochrome oxidase,Tyrosinases, Ascorbate oxidase, Nitrite reductases
Zn^{2+}	Alcohol dehydrogenase, Carbonic anhydrase, DNA polymerase, Pyruvate carboxylase
Mg^{2+}	Glucose 6-phosphatase, Hexokinase, DNA polymerase
Mn^{2+}	Isocitrate dehydrogenase, Arginase
Inorganic elements	Enzymes

Inorganic elements	Enzymes
K+	Pyruvate kinase
Ni^{2+}	Urease, Carbon-monoxide dehydrogenase, Cytochrome-c3 hydrogenase, Phospholipase C
Mo	Nitrate reductase, Nitrogenase
Se	Glutathione peroxidase

Fish is also an important source of vitamins. Vitamins and minerals (micronutrients) together play a vital role in metabolism and keep us healthy. A list of vitamins and minerals and their deficiency diseases is given below. (Table 4)

Table 4: Vitamins and minerals and their deficiency diseases

Vitamins	Deficiency diseases
Vitamin A	Xerophthalmia and night blindness
Vitamin B1 (Thiamine)	Beriberi
Vitamin B2 (Riboflavin)	Non-specific fatigue, eye changes, dermatitis, brain dysfunction
Vitamin B3 (Niacin)	Pellagra
Vitamin B5 (Pantothenic acid)	Rare
Vitamin B6 (Pyridoxine)	Dermatitis, neurological disorders, convulsions, anaemia
Vitamin B7 (Biotin)	Rare
Vitamin B9 (Folate)	Megaloblastic anaemia, neural tube and other birth defects, heart disease
Vitamin B12 (Cyanocobalamin)	Megaloblastic anaemia (Helicobacter pylori induced gastric atrophy)
Vitamin C (Ascorbic acid)	Scurvy
Vitamin D (Dihydrocholecalciferol)	Osteoporosis, Rickets
Minerals	
Iron	Iron deficiency anaemia
Calcium	Osteoporosis
Iodine	Goitre, hypothyroidism
Zinc	Poor pregnancy outcome, impaired growth (stunting), genetic disorders, decreased resistance to infectious diseases
Fluoride	Increased dental decay, affects bone health

Role of fish consumption on material and child health

Maternal health is important for the individual's and the newborn's health. A Mother needs nutrients for her health maintenance and additional nutrients to provide a steady supply of all nutrients (sugars, amino acids, proteins, fatty acids, minerals, and vitamins) for the growing foetus in her womb.

Any deficiency during pregnancy will affect the growth of the foetus and the newborn, leading to developmental defects, complications during pregnancy, low birth weight, and FOAD (Foetal Onset of Adult Diseases).

Fish consumption during pregnancy has been correlated with good pregnancy outcomes and normal, healthy babies. A case study was reported from a population living near Lake Victoria, Africa. It showed that the population with a low social makeup was dependent solely on the fish living in the lake for their livelihood. Fatty acid profiling of the breast milk of feeding mothers showed a similar FA profile to that of the fish consumed from the lake (Fiorella *et al.*, 2017); thus, providing strong evidence that fish consumed during pregnancy has a direct bearing on the growing foetus and the newborn.

Mass awareness programme to generate public perception of health benefits of fish

Keeping all the above nutritional facts in mind, it is strongly recommended to consume fish regularly. The main objective of the mass awareness programme was to bring awareness to the public about the health benefits of fish consumption by explaining the nutritional and scientific facts to them in the common man's language. Numerous mass awareness campaigns were conducted across the nation, including different parts of Odisha, particularly in the districts of Mayurbhanj, Balasore, and Jagatsinghpur, under USAID-World Fish, Odisha-sponsored programmes in which the target audience participated in large numbers.

Fig. 1.1: Mass awareness program at the village Sagunadiha, Mayurbhanj, Odisha

---

Kathryn J. Fiorella, Erin M. Milner, Elizabeth Bukusi and Lia CH Fernald, 2017. Quantity and species of fish consumed shape breast-milk fatty acid concentrations around Lake Victoria, Kenya. Public Health Nutrition: 21(4):1-8 DOI: 10.1017/S1368980017003147 (Accessed on 05-10-2023)

The target audience included women in particular, as well as school children and other members of the public. Producers (farmers) and consumers were explained the benefits of fish consumption and, therefore, the need for taking up small-scale fisheries and aquaculture, as well as commercial farming on large scale.

The concept of 'carp-mola polyculture' was also explained. As big fishes like the carps (rohu, catla, and mrigal) are necessary for ensuring protein supply to a vast population to keep the protein hunger (protein-calorie malnutrition) away, small indigenous fishes are especially important as they are vitamin and mineral-rich (micronutrient-dense) and are necessary for preventing the micronutrient deficiency (or hidden hunger).

Fig. 1.2: Mass awareness program at Balasore, Odisha

Fig. 1.3: Mass awareness program at Jaipur, Jagatsinghapur district, Odisha

2

Empowerment of Rural Tribal Population through Sustainable Inland Fisheries Development in Sagar Island Sundarban, West Bengal

Archana Sinha

ICAR-Central Inland Fisheries Research Institute (ICAR-CIFRI)
Barrackpore, Kolkata, West Bengal

The village of Khan Saheber Abad is located in the Sagar subdivision of South 24 Parganas district in West Bengal, India. It is situated 4.8 km away from the sub-district headquarter Rudranagar (the tehsildar office) and 99.6 km away from the district headquarter Alipore. A training programme was conducted for the empowerment of rural tribal communities through sustainable inland fisheries development and utilisation of open water bodies in the village.

Fig. 2.1: A view of Ganga Sagar and the Kapilmuni Temple at Ganga Sagar

Table 2.1: Summary Information

Location	Village: KhanSaheber Abad	PS: Sagar Tahsil	Dist: South 24 Parganas	State: West Bengal
GPS Coordinates	Latitude: 22° 57' 0" & Longitude: N 90° 19' 0" E			
Population	Total:6082	Male:3118	Female:2964	
Participants	Total:165			

Fig. 2.2: Tribals of the village dancing with Hon'ble Union Minister Madam Uma Bharti ji

Challenges and opportunities

The Khan Saheber Abad is primarily a tribal dominated village. These Tribal people belong to the Santhal community. They have a low level of literacy, declining or stagnant population, and are economically weak, but they are very close to nature. The people of this village used to earn their daily livings by working as labourers. With meagre earnings they faced hardship to meet their daily needs. Fisheries was not a means of livelihood for the villagers. They were catching fish from some of the natural low-lying water bodies. However, providing frequent trainings and demonstrations at the village, has attracted youths, including women to rear indigenously available fish for livelihood.

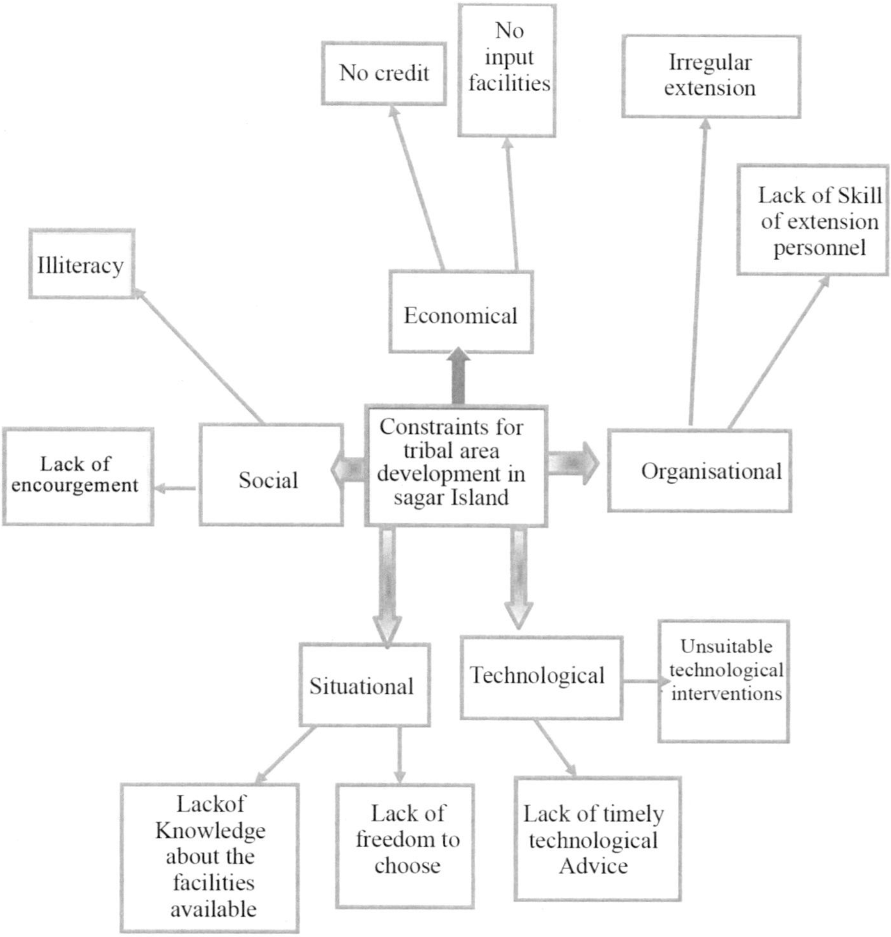

Fig. 2.3: Constrains impeding tribal area development in Sagar Island

Demonstration site

Derelict water bodies and indigenously available fish species were selected for culture:

- Area: 0.667 ha. (Ten units of derelict water bodies of different sizes vary from 0.027 to 0.195 ha)
- Number of farmers: 10 groups, each having four families, with a total of 165 individuals.

Interventions deployed

Scientific interventions, *viz.,*. natural resource identification and inventorization; reducing the cost of fish feed by on-farm feed formulation and azolla culture; integrated fish farming; application of vermicompost; biofloc; and periphyton, were the major initiatives of the programme. Attempts were made to stock the self-recruiting small indigenous fish species along with the Indian major carp species to provide a continuous supply of fish to the farmers throughout the year.

Demonstration and implementation

The members of the tribal community (both male and female) were the target group of the initiative. The target population was below the poverty line; therefore,a livelihood support component was also included in the programme. ICAR- CIFRI, Kolkata Centre, collaborated with Swami Vivekananda Youth Cultural Society, Krishnanagar, Sagar Island, to develop grass roots networking on the island. The programme was launched with a two-day skill development and awareness programme on "Ecology and fish diversity conservation."

By providing frequent trainings, demonstrations, and sensitization both on-site at the village and at the institute we drew the attention of a group of youth, including women, to farm locally available fish for their livelihood. Derelict water bodies were selected to rear the fish. These selected water bodies were cleaned and prepared for fish farming by the villagers under the supervision of scientists from ICAR-CIFRI. The locally available small indigenous fish species, which are self-recruiting species, were stocked along with the Indian major carp for yearround availability of fish to farmers for their consumption as well as for market. The rearing units, which have tidal influx and high salinity, are used for rearing brackishwater fish, *viz. Liza parsia, Rhinomugil corsula, Scatophagus argus*, *Sillago panjijius*, *Liza tade*, *Pseudapocryptes lanceolatus,* and others. Partial harvesting and natural stocking of fish during the monsoon months reduced the burden of seed costs for the farmer. The self-recruiting small indigenous fish species, viz. *Amblypharyngodon mola, Puntis sarana, Salmostoma bacaila*, *Mystus gulio,* etc., were initially stocked during the first year of farming, but being self-recruiting, they were present in the culture system even through the third year. These fish were used by the farmer for their self-consumption, and thus, they played an important role in the food and nutritional security of the farmer, and their families.

Fig. 2.4: Training the beneficiaries

Fig. 2.5: Stocking of fish seed

Support for farmers

A series of skill development and mass awareness programmes were conducted on "Ecology and Fish Diversity Conservation" to sensitise the tribal farmers of Sagar Island. The need-based capacity building programmes were organised at ICAR-CIFRI, Barrackpore, and Kolkata, and the requisite items such as fish seed, fish feed, lime, nets, and water pumps were provided to the farmers under the Institutes Tribal Sub Plan activities, which in turn gave better returns and attracted many more villagers to take up fish farming as a source of food, nutrition, and livelihood.

Fig. 2.6: Training and distribution fish seed and nets to farmers

Outcomes

In the first six months of rearing fishe in three freshwater and two brackish natural water units, the farmer, by partial harvesting, could generate a revenue of Rs.18,599 by selling fish. In the second year, the farmers were more attentive, monitored the process as per the technical support provided, and were able to earn a sum of Rs. 52,865 from the partial harvesting of fish from the same water bodies. Later, based on the request of the villagers five more fish farming units were added, and fish seeds were stocked. The growth rate of stocked fish was monitored and found satisfactory. Among the Indian major carps stocked, the catla attained a size of 700-800 g in six months. The demand for *Amblypharygodon mola* was considerably high and was utilised by the locals in their diet.

A high return with a cost-benefit ratio of 1:3 made villagers happy to adopt fisheries as an alternative source of livelihood. Ten families and about 165 individuals benefitted from this programme and were able to earn more than Rs. 2 lakh annually.

Impacts

Villagers and panchayat members are now interested in adopting fisheries in water bodies lying unused. Even senior citizens of the village are showing interest in fish culture for additional income and livelihood. Almost all families are able to eat fish as well as earn money by selling fish. Shri Kanai Hembarg, President of Adivashi Yuvak Sangh, has acknowledged that he was able to earn Rs. 40,000 by selling fish. The rearing of small indigenous self-recruiting fish species along with conventional carp generated additional income and nutritional security for the poor tribal families. The knowledge that they should eat small fish for their health has given them a way to rear small indigenous fish in small natural water bodies.

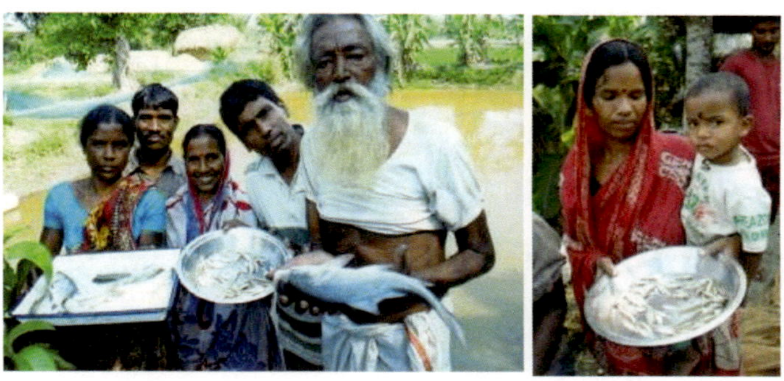

Fig. 2.7: Joyful members of the family with healthy fish

Fig. 2.8: Fish harvest from beneficiaries' ponds in the presence of Hon'ble Sri Bankim Chandra Hazra, MLA, Sagar Island.

3

Empowerment of Fish Farmers Through Sustainable Canal Fisheries Development in Frazergunj, Bakkhali Sundarban, West Bengal

Archana Sinha

ICAR-Central Inland Fisheries. Research Institute (ICAR-CIFRI) Barrackpore, Kolkata, West Bengal

Canals and waterways are the most important source for irrigation and have been identified as one of the most important aquatic resources for fish culture. The natural canal systems in the Sundarbans cover an area of 907.33 ha. Fisheries activities were planned in order to provide a source of livelihood to the underprivileged people of the cyclone-prone area of the Sundarbans.

Table 3.1: Summary information sheet

Location	Village: Shibpur	PS: Frasergunj	District: South 24 Parganas	State: West Bengal
GPS Coordinates	Latitude: (21°36’39.3”N Longitude: 88°15’21.6”E)			
Population	Total:9303	Male:4735	Female:4568	
Participants	Total:100	Male: 80	Female:20	

Challenges and opportunities

Fisheries in the Sundarbans are poorly documented and less attention is given to producing fish from canals. Information on the trophic status of the canal habitat, biotic community structure, and hydrological parameters is also lacking. However, the villagers reported that the fish enter these canals from the source waters naturally, and form natural populations. The villagers residing beside the canal catch fish, especially during monsoon, and the women of the adjoining households often collect small indigenous fish (SIFs) or prawns through small dragnets or scoop nets for household consumption. The shared information of the villagers was utilised to sensitise them and convince them to initiate fish culture in these canals.

Demonstration site

To utilise these unexplored aquatic resources, a group of fish farmers was formed. They were sensitised, and a viable method of fish culture was demonstrated in Bharua canal, Sundarbans, West Bengal, with community participation. The canal is located in Shibpur, Frasergunj (21°3639.3N, 88°1521.6E) and is tidal fed, and connected to the Hoogly River. The total length of the canal is 2 km and its width is approximately 45 m with a water depth ranging from three to eight feet.

Interventions

Applied novelty to culture fish in a modified enclosure system called Net Partition system with community participation.

Fig. 3.1: PRA was conducted to evaluate the method of implementation of the programme

Fig. 3.2: Net partition in a canal as a type of enclosure system

Demonstration and implementation

A Memorandum of Understanding (MoU) was signed between the local stakeholders and ICAR-CIFRI, Barrackpore, regarding the culture of fish in the selected canal. In collaboration with the beneficiaries, a net screen barrier as a partition system was installed in the Bharua canal. The partitions were made with locally available bamboo poles, HDPE net screens, and mosquito nets to reduce the cost. Three net partitions were constructed, with the dimensions of each partition being 50m × 45m (L × B), covering a total area of 2,250 sq. m. Healthy fingerlings of Indian major carps, i.e., *L. catla*, *L. rohita,* and the exotic carp *C. idella,* with an average weight of 5.84±0.10g, 5.65±0.48g, and 6.49±0.26g respectively, were obtained from a fish seed supplier at Naihati, North 24 Parganas, Kolkata, West Bengal and stocked. Before stocking, fish were acclimatised in hapa for one week. Fish were stocked in a net partition system at 2 nos./sq.m with a stocking ratio of 50:40:10, respectively. The fish were fed with a commercial feed containing 28% crude protein and 5% crude lipid twice a day (10.00 hrs. and 16.00 hrs.). After stocking, fish were reared for 180 days.

Fig. 3.3: Field demonstration in the canal

Fig. 3.4: Training conducted at the site for knowledge development

The maximum size recorded after 180 days of culture for catla, rohu, and mrigal was 1.2 kg, 1.0 kg, and 0.80 kg, respectively. This encouraged the local community to culture fish in canals. Apart from IMC, a good quantity of small indigenous fish (SIF's) like *Puntius sophore*, *Puntius ticto, Amblypharngyodon mola, Salmostoma bacaila,* and freshwater prawn *Macrobrachium rosenbergii* were also harvested. Harvested fish were sold in the local market at a higher price. All this encouraged the local villagers to culture fish in the canals of the Sundarbans. Demonstration of cultural practices with diversified species is advised in these canals and perennial irrigation canals that exist in different states. The use of systematic net partition systems, covering large areas withlow-investment, can also be an option for enabling fish production and livelihood support for the local rural people.

Fig. 3.5: Fish harvest from Bharua Canal, Sundarbans

Table 3.2: Economics of IMC and EC rearing in the Net Partition System in the Bharua Canal

Particulars	Rate (₹)	Unit	Amount (₹)
Capital costs			
Cost of bamboo poles (number)	300	45	13500
Cost of HDPE net screen (metre)	400	50	20000
A. Subtotal			**33500**
B. Capital costs per crop (assuming that the materials will last for 3 crops)			11166.67
Operational costs			
Cost of fingerlings (kg)	200	27	5400
Cost of Feed (kg)	32.5	2000	65000
Labour charges for the construction of NPS	350	7	2450
C. Sub-total			**72850**
Total costs (B+C)			**84016.67**
Gross income			
Sale value or price of fish			
C. catla (kg)	120	463.34	55600.80
L. rohita (kg)	120	275.34	33040.80
C. idella (kg)	100	89.44	8944

Particulars	Rate (₹)	Unit	Amount (₹)
D. Sub-total			**97585.60**
E. Net income (gross income - total **costs)**			13568.93
Benefit-cost ratio	1.16		
Return on investment	0.16		

Impacts

There is a conflict of interest between agriculture and fisheries as most of the canals are used for irrigation purposes. Governance is also a major issue, as these are common property resources. At the same time, the probable impact on biodiversity and ecological integrity of these resources is also a concern, which needs more cutting-edge research on these resources. For successful management and development in this sector, it is essential to have an appropriate understanding of these ecosystems with site-specific, suitable methodologies and enhancement practices for fish production and plausible guidelines, which can maintain the role of canal fisheries as a social "safety net" along with the promotion of fishers in the locality. The canal system in the Sundarbans holds great potential for fish production, but for a long time it has been underutilised or unutilised. The canal is a source of livelihood for the local population and also mitigates their nutritional needs. Moreover, the present study showed that the culture of fish in a net partition system in a canal is a feasible option. Adoption of a net partition system in a canal system can provide an alternate livelihood option to the fishers' community, as the Indian major carp has good demand in local markets in addition to the native fish of the canal.

Fig. 3.6: A village boy with prawn

4

Conservation of Small Indigenous Fishes with Community Participation for Nutrition and Livelihood Security

[1]Archana Sinha and [2]Bimal Prasanna Mohanty

[1]ICAR-Central Inland Fisheries. Research Institute (ICAR-CIFRI) Barrackpore, Kolkata, West Bengal
[2]Indian Council of Agricultural Research (ICAR), New Delhi, India

Village Madangunj is situated on the bank of the Hatania Duaniya, a branch of the Hoogly River in Namkhana. The cyclone *Aila* unleashed its fury on the people residing on the islands in the Sundarbans and changed their lives forever. The Sundarban is the world's largest mangrove forest and home to the Royal Bengal Tiger. The cyclone also affected the fish resources and livelihood of the villagers. Efforts were made to provide alternative sources of livelihood to the poor people of these villages.

Fig. 4.1: River survey in Namkhana, Sunderbans

Challenges and opportunities

We conducted field surveys to understand the status of small indigenous fish after cyclone *Aila* in the Sunderban region, particularly in Namkhana block, which was badly affected by the cyclone. Hundreds of people under various age groups ranging from 27 to 65 years old and involved in various occupations were interviewed personally to know about the present status of small indigenous fish after *Aila*. It was reported by 60% of the respondents that there is a significant reduction in the availability of small indigenous fish (SIF). To rejuvenate their stocks, it was decided to develop a conservation site, as a common seed bank for SIF. The SIF are self-recruiting species and can be cultured in various cultural systems like ponds, canals, wetlands, cages, etc. They can tolerate the adverse water quality conditions and have a higher growth and maturity rate.

Demonstration site

Village Madangunj of Namkhana comes under the subdivision Kakdwip in the district of South 24 Parganas in West Bengal, India. Namkhana Block, with an area of 370.61 sq km, lies in the south-western corner of West Bengal. It is connected with Kolkata by an overall 132 km long all-weather road and also by a ferry service over 150 m of water separating the island, which has a length of 27 km from the mainland.

Interventions

An initiative was taken to conserve selected locally available SIF species in the Sunderban area. The conservation and demonstration sites were developed in the village community pond of Madanganj, Namkhana. Fish were collected locally from the rain fed canals and stocked. Two farmer interest groups (FIGs) were formed, each with six members. The groups were monitored by the Madanganj Matsya Chasi Samiti, Madanganj, Namkhana.

Fig. 4.2: A meeting was conducted with the Sundarban Development Board, WBUA&FS, Panchyat, and Block members to develop the conservation site for SIFs.

Mass awaranes and training programmes organised

The mass awareness drive was conducted to spread awareness in rural communities regarding small indigenous fish (SIF) for nutritional as well as livelihood security. A multidisciplinary team of scientists from ICAR-CIFRI, Barrckpore, took the initiative to conduct this mass awareness camp in association with Panchayat Samiti and Madanganj Cooperative Society. The opportunity was utilised to educate the fishermen and women regarding the nutritional benefits of the small fish that are locally available. The importance of conservation of small indigenous fish was also demonstrated during the programme.

Fig. 4.3: Organised a mass awareness programme on "Small Indigenous Fish for Nutritional Security and Rural Livelihood" on the occasion of World

Fig. 4.4: Capacity building programme on "Importance of Small Indigenous fish for Livelihood and Nutritional Security".

Fig. 4.5: Training of beneficiaries in the Institute

Demonstration and implementation of the process

SIF, namely Koi *(Anabus testudineus)*, Tangra *(Mystus vittatus)*, Mourala *(Amblypharyngodon mola)*, Punti *(Puntius ticto)*, Bogo *(Xenentodon cancila)*, Bele *(Glossogobius giuris)*, Pankal *(Mastacembelus pancalus)*, Lata *(Channa punctatus)*, and Chanda *(Chanda nama, Chanda ranga)* collected from the canal of Madanganj area, were stocked in the village community pond. As they are self-recruiting species, a significant increase in their numbers was recorded.

Fig. 4.6: Fish Conservation Site at Madangunj (size 165’x125’)

Fig. 4.7: Small fish caught from the rain- fed canal at Madanganj

Fig. 4.8: Harvest of SIF from the site

5

Hygienic Handling, Packaging Branding, and Quality Certification of Fresh and Dry Fish

Sushrirekha Das and Shahaji Phand

National Institute of Agricultural Extension Management, MANAGE Hyderabad

Fresh fish serves as a protein-rich food for people of all income groups. Dry fish is in greater demand when the availability of fresh fish in the market is low. Dry fish have a higher concentration of protein (in terms of weight) as compared to wet fish and are an inexpensive source of animal protein. Moreover, dry fish production provides employment opportunities to fishermen, fish farmers, rural youth, and women and also helps them to generate extra income. In addition, it also provides food and nutritional security to the rural, underprivileged population. Though dry fish enterprises have a significant role in improving the livelihood as well as the nutritional security of society, not much attention has been paid to the production and marketing of dry fish products. Lack of awareness about the requirements of hygiene, packaging, and quality certifications, coupled with the limitations of infrastructure has often resulted in heavy losses. The high incidence of post-harvest losses is a major impediment to the realisation of the government's goal of increasing the contribution of this sector to the overall national economy.

The training programme was sponsored by NFDB-MANAGE and conducted by the Department of Fish Processing Technology, Dr. MGR FCRI, on "Hygienic Production, Packing, Branding, and Quality Certification of Dry Fish" on 17, December 2021, under the leadership of Dr. Balasundari, Dean, Dr. MGR FCRI. Dr. Shahaji Phand, Centre Head, EAAS, MANAGE coordinator, and Dr. Sushrirekha Das, MANAGE Fellow, co-coordinator of the programme. There are 50 participants in the program, who were coastal fishermen. Out of 50 the participants, 45 were women, and 5 were men. Their day- to- day earnings depend on the fish they catch at sea, and it plays a very crucial role in socio-economic development, it increases revenue, enhances employment opportunities among the Scheduled Caste Fishermen, Farmers, Youth, and Women, and supplies superior quality dry fish.

The trainees shared their feedback and were assured to take up the dry fish production business very shortly.

Training objectives

- To create awareness among the Scheduled Caste Fishermen, Farmers, Youth, and Women on "Hygienic Production, Packing, Branding, and Quality Certification of Fresh and Dry Fish."
- To provide easy and low-cost AIG options for coastal fishing communities to improve livelihood.
- To identify the reasons for fish spoilage, quality loss, and associated health risks.
- Low-cost processing can effectively prevent spoilage, reduce post-harvest fishery losses, and improve seafood safety.
- To describe the steps in carrying out sun drying and fish salting, paying particular attention.

To ensure adequate handling of raw material fish during preparation, processing, storage and marketing.

Fig. 5.1: A view of training programmes organised

Demonstration

The trainer has taught the participants about the process of hygienic production, packing, branding, and quality certification of fresh and dry fish with practical demonstrations. The steps were presented in a flow chart.

Raw Fish

↓

Sun-drying of large-sized fish,elongated fish,small fish, and shrimp

↓

Dressing/gutting/washing

↓

Sun-drying for 35 days

↓

Cooling and storage in jute bags

Flow chart for the traditional drying process for marine fish in the coastal region of Tamil Nadu:

Fig. 5.2: Demonstration of preparation methods by the trainer to participants

The importance of the training programmes

1. Knowledge gained from the training

In the water, there are millions of microscopic organisms known as bacteria that can come into direct contact with fish. During respiration, water travels via the mouth and gills. These bacteria may enter the stomachs of marine fish because they sip water. As a result, microorganisms are exposed to the entire body of the fish (skin, eyes, mouth, gills, and stomach). The majority of these bacteria are harmless: some are beneficial to fish, while others break down plants in the soil to release nutrients into the water. However, certain bacteria cause disease in fish and ruin them. When the fish is alive, however, the spoilage bacteria cannot ruin the muscle. After death, spoilage bacteria enter the muscle through the skin and gills, breakdown the muscle cells, and use the energy required for growth.

The supply of energy stops when a fish dies. Within a few hours of death, the fish's muscles tighten and the entire body stiffens. Depending on the species, temperature, and other factors at the time of death, this hardening can last for a few hours. As the muscle produces more acids, spoilage bacteria from the surface, gills, and gut attack and feed on the smaller components, spreading out. Increased bacterial activity causes a thick slime to form on the skin and gills, as well as a sour odour and softened muscular tissue.

2. Value addition

Farmers suffer losses in dry fish processing due to very little knowledge of the processing part. In this connection, the training covered the entire process of dry fish processing. The major parts are the sun drying process, salting method, packaging and storing of dry fish, and control of beetles and mites. The training feedback has been given by the participants. Out of 54 participants, 52 gave a score out of 10 to 10, and the remaining two participants scored 8, as they were absent from one class, which was cleared later. The training participants, especially the female participants, were very happy to learn the training process of the dry fish business, which they can utilise in their day-to-day-to-dry fish. They realised their lacuna, where they were missing the process. All the participants promised to be good entrepreneurs in the processing, hygienic production, packing, branding, and quality of dry fish.

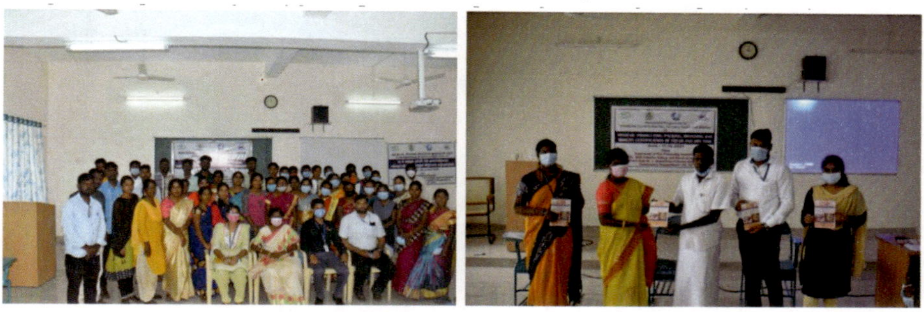

Fig. 5.3: Learning and input distribution

Impacts

The training carried out by MGR Fishery College completed the programme nicely. The selected training topics were taught in the theory classes and are better understood by the participants. Almost all the processes in the hygienic production, packing, branding, and quality of dry fish were learned in theory as well as in practice. More so, the training programme gave the entrepreneur a platform to generate income in the field of dry fish processing.

6

Water Treatment and Disease Management in the Biofloc System

Sushrirekha Das[1], Nityasundar Pal[2] and Shahaji Phand[1]

[1]National Institute of Agricultural Extension Management (MANAGE) Hyderabad
[2]National Fisheries Development Board, NFDB, Hyderabad

The demand for nutrients is rising, along with the population, which is rapidly expanding. The sustainability of life on earth now places a high focus on protecting the sources of sustenance. In this regard, aquaculture has a crucial role to play. The regulated raising of fish and other breeds in a particular environment is known as aquaculture. The natural resources and other elements required for sustainable development can be preserved using this method. The promotion of the value of nutrients and the expansion of employment prospects in the future are both made possible by aquaculture. Biofloc fish farming, which has set trends in fish farming all over India, facilitates recycling and reusing of the nutrients in the culture medium by filtering the unwanted substances and metabolites and converting them into nutrients, which benefits the zero exchange of water. The major fact of this farming system is that it reduces feed intake, which leads to less investment in protein-rich feed, which costs a lot.

Fig. 6.1: A Biofloc Unit set up at Jagatsighpur, Odisha, under PMMSY.

Biofloc training in Balasore and Jagatsinghpur, Odisha

Due to the poorly known dynamics of this technology, there is a significant risk that even a small amount of carelessness will have a big negative impact on the lives in the pond. Aqua One Centre in Balasore, Odisha, is an aquaculture service provider company that offers a wide range of services to support the development and growth of the aquaculture industry, mainly in biofloc based fish farming.

It is great to see that young male and female entrepreneurs in Odisha are showing interest in biofloc fish farming and that training programmes are being organised to help them learn the necessary skills and techniques. The efforts of Mr. Mahidhar Bolem and the Marsco Aqua Clinics team in organising these training programmes is commendable. It is also encouraging to see that the training programmes have attracted participants from other states as well, indicating a growing interest in sustainable aquaculture practices across the country. The fact that training programmes were continued even during the COVID-19 pandemic through online sessions is also commendable. This shows the commitment of the trainers to ensure that the participants receive the necessary training, even in challenging circumstances. Overall, these training programmes are helping to promote sustainable aquaculture practices and support the growth of the aquaculture industry in Odisha and beyond.

Fig. 6.2: Training at Balasore and Jagatsinghpur, Odisha

Table 6.1: Biofloc training conducted by Marsco Aqua Clinics, AOC

Training Batch	Dates	Training Venue	Physical Mode/ Online Mode	No. of Beneficiaries		
				Male	Female	Total
1	26-27th Feb 20	Balasore	Physical	18	12	30
2	24-25 June 20	-	Online (Webex)	17	01	18
3	27-28th July 20	-	Online (Webex)	12	03	15
4	20-21st Aug 20	-	Online (Webex)	12	10	22
5	2-3rd Sept 20	-	Online (Webex)	17	02	19

Training Batch	Dates	Training Venue	Physical Mode/ Online Mode	No. of Beneficiaries		
				Male	Female	Total
6	21-22nd Nov 20	Balasore	Physical	12	08	20
7	17-18th Dec 20	Balasore	Physical	15	05	20
8	22-23rd Jan 21	Balasore	Physical	18	09	27
9	20-21st Feb 21	Jagatsinghpur	Physical	15	05	20
10	26-27th Feb 21	Balasore	Physical	10	0	10
11	28-29th March 21	Jagatsinghpur	Physical	13	02	15
12	16-17 April 21	-	Online (Webex)	12	0	12
13	21-22 April 21	-	Online (Webex)	10	0	10
14	9th July 2021	Jagatsinghpur	Physical	13	02	15
Total				**194**	**59**	**253**

NFDB-MANAGE sponsored SCSP training programme

The National Institute of Agricultural Management (MANAGE), Hyderabad, and the National Fisheries Development Board (NFDB), Hyderabad, have sponsored a 5-day training programme on biofloc pre- and post water treatment and disease management programme. The programme was conducted by Marsco Aqua Clinics, AOC, under the guidance of the managing director. It is also encouraging to note that the training programme had technical sessions handled by experts from NFDB, BBSR, the State Fisheries Office, and Marsco scientists. This indicates that the training programme was comprehensive and provided participants with a wide range of knowledge and skills.

It is also noteworthy that the training programme included four field visits with demonstrations, which would have allowed participants to see first-hand the application of the techniques and practices they learned during the technical sessions.

The fact that the training program covered topics such as the generation of biofloc,pre-stocking, post-stocking, feeding, management, and economics of biofloc farming activities is noteworthy. This indicates that the training programme was designed to provide participants with a holistic understanding of the subject and equip them with the knowledge and skills needed to successfully practice biofloc farming.

Overall, such comprehensive training programmes are essential for promoting sustainable aquaculture practices and supporting the growth of the aquaculture industry in the region. It is commendable that Marsco Aqua Clinics, AOC and the other organisations involved in the training programme took the initiative to provide such training opportunities to aspiring aquaculturists.

Fig. 6.3: Training programme organised at Jagatsinghpur, Odisha

It is also noteworthy that students attended the programme with great enthusiasm, indicating that there is a growing interest in sustainable aquaculture practices among the younger generation. The fact that 31 male and 27 female beneficiaries participated in the programme is also noteworthy, as it indicates gender-balanced participation.

It is encouraging to learn about another training programme on biofloc based culture technology that was organised in Jajpur, Odisha. It is noteworthy that the programme was designed to be inclusive, with 50 beneficiaries from the Schedule Caste (SC) community participating in the training.

It is also commendable that the programme covered both theoretical and practical sessions, and that there were field visits with demonstrations to provide hands-on training to the participants.

The discussion on converting toxic material such as fish excreta into single-cell protein and the use of fermented carbon organic (FCO) to eliminate the use of pelleted feed is noteworthy, as it highlights the importance of sustainable and cost-effective practices in aquaculture. It is also encouraging to learn about the subsidy scheme facility for farmers, and the fact that the Odisha government provided a subsidy of 40% for general category beneficiaries, with an increased subsidy of 60% for female and SC/ST beneficiaries. Such initiatives can play an important role in promoting inclusivity and diversity in the aquaculture industry and supporting its sustainable growth.

Fig. 6.4: MANAGE, Hyderabad, conducted the training through the WebEx platform

Training objectives

- To promote biofloc farming technology among the rural unemployed youth for livelihood.
- To make biofloc farming technology a more commercially viable option.
- To educate the farmers about modern and scientific methods of biofloc technology through training.
- To develop new entrepreneurs in the biofloc sector.

Fig. 6.5: Practical session cum field visit tour of participants

Is biofloc based farming profitable?

The trainers explained to the participants the profitability of biofloc-based farming depends on several factors, such as the cost of inputs, the market demand for fish, the selling price, and the efficiency of the farming system. However, when done properly, biofloc-based farming has been shown to be a profitable venture. The use of biofloc technology can significantly reduce the cost of feed, which is typically the most expensive component of fish farming. Additionally, biofloc-based farming can result in higher yields of fish per unit area of water as compared to traditional pond culture methods. This increased productivity can lead to higher profits. However, it is important to note that proper management practices are necessary for the success of biofloc-based farming, and farmers must ensure that they follow all guidelines and recommendations to ensure profitability.

In biofloc farming, there are certain aspects that make investment low and certain things that are pricey. There are currently more meat eaters than vegetarians in the world. Additionally, non-vegetarian foods provide the body with the required nutrients. Non-vegetarian Males make up 80% of the made population of India, whereas non-vegetarian females make up 60%. Of these, fish is preferred by 48% of the men and 42% of the women. To meet the demand for fish in their daily diets, these enormous populations require a lot of fish in the market. Because of this, both small and large fish can greatly benefit from biofloc fish farming.

According to the quantity of tanks, it requires a specific amount of land. For one tank, we'll need a 15x15ft area. In a 1 kattha area, we can set up 8 tanks. When it comes to tank prices,one tarpaulin tank with a metal frame and outlet costs around 20,000-25,000 INR. Cement tanks are only 5,000 INR expensive than tarpaulin tanks. However, because these tanks last for many years, this investment appears to be a good long-term investment. Other items that require expenditure include an oxygen machine, pipes, 24-hour electricity, and probiotics. Another investment is in the purchase of baby fish (seed) and their feed.

Biofloc fish farming can yield a significant return if done correctly. Patience is required for successful fish aquaculture. If all goes well, we should be able to earn a return on investment in less than a year. Biofloc farming allows us to harvest twice every year. The fish weigh between 500 and 800 grams. 1 kg of rohu orcatla fish costs about Rs. 150 to purchase in market. If the farmer sells Rs.150 in his orher local market, he/she can definitely earn 30-40 Rs. per kg, and depending on the fish breed and market demand, this profit could be more. As a result of this information, it is evident that Biofloc is a profitable business. However, if done without sufficient research, training, or technical knowledge, it could result in significant losses.

The participants found the training programme useful and believe that the biofloc technique can help double their income. The use of natural bacterial and plankton stock through biofloc technology can certainly reduce the reliance on expensive pelleted feed and improve the overall health and growth of the fish. We hope that the farmers who have attended the training will be able to apply the techniques they learned to their own farms and achieve success in their aquaculture businesses.

Fig. 6.6: Group photo of the participants

Table 6.2: Participants attending the training programmes

Area	**No. of Beneficiaries**		
	Male	Female	Total
Jagatsinghpur	29	01	30
Jajpur	20	30	50
Total	**49**	**31**	**80**

Impact

It's always good to see that training programmes are making a positive impact on participants and their communities. By providing training and support, organisations like Marsco Aqua Clinics and NFDB are helping to promote sustainable aquaculture practices and improve the livelihoods of small-scale farmers. It's also encouraging to see that the participants are applying for subsidies and using the knowledge gained in the training to improve their farming practices and increase their profits. Biofloc fish farming has helped farmers and entrepreneurs improve their livelihoods through sustainable aquaculture practices. This technique of fish farming can indeed be a cost-effective and profitable venture, as it reduces the major costs involved in traditional fish farming methods. Overall, initiatives like these can help boost the economic growth of the aquaculture industry in India and empower farmers to become successful aqua entrepreneurs.

Fig. 6.7: A successful Biofloc farmer

7

The Importance of Soil and Water Testing in Aquafarming

Sushrirekha Das[1], Nityasundar Pal[2] and Shahaji Phand[1]

[1]National Institute of Agricultural Extension Management (MANAGE) Hyderabad, Telangana

[2]National Fisheries Development Board, NFDB, Hyderabad, Telangana

It is clear that Odisha has significant potential for aquaculture due to its vast brackish water resources. However, as the aquaculture industry grows, it is important for farmers to be knowledgeable about the inputs and techniques they use to ensure the sustainability of the industry. A total of 2.98 lakh ha of estuaries, 0.081 lakh ha of backwater, 0.80 lakh ha of Chilika Lake, and 0.32 lakh ha of cultivable brackish water area make up the 4.18 lakh ha potential brackish water area for fish and shrimp production.

Additionally, proper disease diagnosis and management are crucial for reducing the risk of disease outbreaks and ensuring the health of the aquatic animals being farmed. With appropriate training and support from organisations such as the Aqua One Centre, farmers can improve their practices and increase their yields, while also contributing to the economic growth of the region.

Fig. 7.1: Map showing Coastal Districts of Odisha

Training on- laboratory testing of soil and water quality for fish/ shrimp farming

The Aqua One Centre, AOC, and Balasore organised multiple training and awareness programmes in five coastal districts of Odisha, including Jagatsinghpur, Balasore, Kendrapara, Bhadrak, and Puri. It's also commendable that the organisation has conducted a total of 45 such programmes and 684 farmers participated in these programmes, with 680 being men and 4 being women. Along with Suryo Udyog Pvt. Ltd. and Solidaridad Asia, they also conducted 5 training programmes in Balasore and Bhadrak districts. These programmes focused on better management practices for shrimp and fish farming and the importance of soil and water testing in aquaculture farming. Overall, these efforts are expected to have a positive impact on the aquaculture industry in the region by educating farmers and promoting sustainable practices.

Fig. 7.2: Training sessions in progress.

One day training programme for beneficiaries

It is important for farmers to test the water in their ponds on a regular basis to understand the basic parameters of the water, such as alkalinity, salinity, and pH. Without this knowledge, it can be difficult to know how much medicine or chemicals to use when problems arise. Farmers always rely on medicines or chemicals when they encounter any kind of problem in their ponds. Regular water testing is necessary to understand the major problems in the pond and to determine the appropriate remedies, such as the right amount of medicine or chemicals to use. It is recommended that farmers test the water in their ponds at least once every 15 days during the rainy and winter seasons.

Fig. 7.3: Demonstration and training of water and soil testing

NFDB-MANAGE sponsored SCSP training

Marsco Aqua Clinics organised a 3-day training programme on the importance of lab testing in aquaculture to improve the culture cycle of shrimp and fish. The training was funded by MANAGE, Hyderabad. With 50 participants, 22 male and 28 female fisher farmers, the programme provided hands-on experience about water and soil testing to facilitate learning. The materials distributed among the participants also added value to the training. The main highlight was the importance of soil, water, and seed testing in both shrimp and fish farming. With soil testing, farmers can determine the amount of lime they need to use before water is poured into the pond, and it can also help detect EHP +ve and ve. Additionally, it was heartening to know that 28 women beneficiaries from the SC community participated in the programme and got clarity regarding the importance of lab testing, which they can utilise in their present culture. Demonstrating how to do the tests on the farm is a great way to ensure that farmers can apply the knowledge gained from the programme effectively.

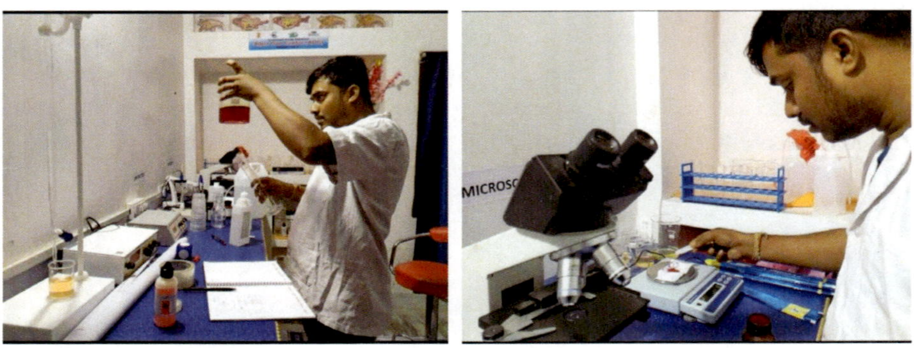

Fig. 7.4: Laboratory testing of soil and water samples

Training and demonstration at the farm

It's important for farmers to regularly test their water quality to ensure optimal conditions for their aquatic species. Water quality factors such as temperature, salinity, dissolved oxygen, alkalinity, hardness, dissolved phosphorus, total ammonia nitrogen, and nitrite are all crucial parameters to monitor. However, only 10% of farmers conduct lab testing for water quality, while the remaining 90% rely on their own practical analysis in the field. This is a potential problem because farmers may not have the necessary tools to accurately assess water quality, leading to suboptimal conditions for their aquatic species. Marsco Aqua Clinics recognised this issue and organised a training programme to raise awareness of the importance of lab testing, in collaboration with MANAGE, Hyderabad. The programme covered various aspects of aquaculture, including hatchery management, common disease problems, and water and soil quality assessment.

The demonstration and hands-on experiments that were shown to the participants during the exposure-visit-cum-training programme were as follows:

- Field training on the accumulation of data for the different parameters required for estimation of disease diagnosis.
- Demonstration of how to test the physio-chemical analysis of pond water and groundwater samples and soil samples like pH, EC, TDS, and salinity.
- Spot sampling methods for collecting water samples from ponds.

Fig. 7.5: Input distribution and farm management

The importance of training

MANAGE, Hyderabad, has allocated a theme to Marsco Aqua Clinics to raise awareness about the importance of lab testing in shrimp farming in Balasore district. Many farmers are not aware of the importance of lab testing, and it can lead to significant losses in farming. The practice of four cycles of shrimp farming in a year can also increase the use of chemicals, which can be harmful to the environment and affect the quality of the product. By raising awareness through training and education programmes, farmers can learn better farming practices, including the use of lab testing, to improve the health and productivity of their farms.

Fig. 7.6: Group of beneficiaries

Discussion about the establishment of aqua clinics

There are various schemes and subsidies available to support the establishment of disease diagnosis and quality testing laboratories in aquaculture under the Pradhan Mantri Matsya Sampada Yojana (PMMSY). The Aqua Clinics under the ACADP can provide support services like pond monitoring, input management, health diagnosis, water analysis, and advisory services to fish farmers, seed growers, etc. ICT tools are poised to play a significant role in developing the fisheries sector in the country.

Having access to Aqua One centres and private laboratories is a positive development for the aquaculture industry in Odisha. However, it is crucial to educate and motivate farmers to utilise these facilities for their own benefit. When farmers start using laboratory testing regularly, it can lead to better production management and increased profits due to reduced production costs. Therefore, it is important to raise awareness among farmers about the benefits of laboratory testing and encourage them to take advantage of the available facilities.

Fig. 7.7: Group of beneficiaries in action

Impact

The trainings were a great understanding platform for the people. It's great to hear that the training programmes were successful in making farmers aware of the importance of soil and water testing in aquaculture farming. It is encouraging to know that 60% of farmers are now collecting water and soil samples at regular intervals for testing. This is a positive step towards adopting scientific aquaculture practices and ensuring sustainable development in the long run. The trained farmers are also reaching out to the trainers for guidance and support when needed. Overall, it seems that the training programmes have been effective in enhancing the knowledge and skills of the farmers and promoting better practices in aquaculture.

8

Opportunities for Fish Farmer Producer Organisations and Stakeholders under Government Schemes

Sushrirekha Das and Shahaji Phand

National Institute of Agricultural Extension Management, (MANAGE) Hyderabad, Telangana

The one-day SCSP-training program on "Opportunities for fish farmer producer companies and government schemes for fisheries stockholders" was organised by Shashwat Sheti Vikas Pratishthan, Shivajinagar, Pune, Maharastra, on 4 January 2022. The training programme was sponsored by MANAGE, Hyderabad. 50 people participated in the training program, consisting of agri-farmers, fishermen and unemployed youths. Out of the 50 participants, 19 were women and 31 were men.

Fig. 8.1: Inauguration of the training programme

The session is all about FPO formation

MANAGE organised a training programme to sensitise farmers about Farmer Producer Organizations (FPOs) and various schemes under PMMSY. FPOs can help farmers take advantage of economies of scale in production and marketing, and the government's plan to establish 10,000 new FPOs by 2024 is a step in the right direction. It is while social media is a powerful tool for spreading information quickly, not everyone has access to it, especially in rural areas. So, training programmes like this can be very effective in reaching out to farmers and providing them with the information they need to make informed decisions about their farming practices.

Farmers, both men and women, have varied gender experiences and encounter different gender limits and opportunities, which must be acknowledged in the rules. Women farmers, for example, may have less access to knowledge, time control, mobility, and even money. Because patriarchal traditions often prohibit women from handling money, women in FPOs may require a different training approach or duration; alternatively, women in FPOs may require extra support to obtain share capital or working capital. Unfortunately, there is no distinct approach for reaching out to women farmers in the guidelines. The programme is gender-blind to the realities of women farmers since it treats men and women equally in an unequal society.

The majority of trainees were youth and women; this effectively indicates that while women farmers are active in low-value agricultural production operations, they will be excluded from the value-added portion of the supply chain, which will be dominated by men. With technological developments, especially in historically female-dominated jobs like grading, sorting, and packaging, women are beginning to be marginalised, and their work is being taken over by men. Women are likely to lose out at the entrepreneurial level if FPOs become the basic unit for value chain and supply chain related schemes in the fishery sector, and their labour will be relegated to the fields.

Fig. 8.2: Farm demonstrations

Training covered the main topics

- Working together versus individual action.
- Basic concepts of an FPO.
- Structure, formation, and functions of FIGs and FPOs.
- **About PMMSY-** By establishing economies of scale, fishers and fish farmers will be economically empowered and their bargaining strength will increase. To improve productivity through utilising resources in a more efficient, cost-effective manner, better liquidity and remunerative market linkages for fishermen and fish producers would result in higher profits. It will also enhance the ability of fishermen and fish growers to build entrepreneurial abilities for the future.
- **Production and Productivity:** Supply of quality inputs like seed, fingerlings, brood stock, fish feed, fishing nets, and such other inputs for production at reasonable rates; undertaking Pond Culture, Pen Culture, Cage Culture, RAS, Raceways, Biofloc etc. related fish culture activities for both inland and marine regions; dissemination of technology, quality control, and other fisheries related activities and innovations; Undertake aggregation of smaller lots of farmer-members produce.
- **Post-Harvest Management and Infrastructure:** Need-based production and post-production machinery and equipment like storage – ice flakes, ice boxes, and transportation/logistic support – reefer vans, insulated cargo and such other machinery and equipment on custom hiring basis for members to reduce the per unit production cost; e cleaning, assaying. sorting, grading, packing and also fish farm level processing facilities at user charge basis at a reasonably cheaper rate; higher income generating activities like cold chain development, seed/

brood stock production, ornamental fisheries, seaweed cultivation, cold water fisheries, fish kiosks, aquarium manufacturing etc.; any activity (including but not limited to microfinance, e-market, technical support, repairs and maintenance services for boats, motors, cold chain, reefer transport etc.) associated with fisheries supply chain as door-step support.

- **Marketing and Branding:** Branding, packaging, labelling, standardisation of products; facilitating market information about the produce for educated decision-making in production and marketing; undertaking the operation of fish vending kiosks at various urban centres; developing fish and fisheries- related products and by-products; and tie ups for domestic and export sales.

Fig. 8.3: Training session conducted by expert SSV, Pune

Outcome of the training programme

Training programme mean to help the women and fishermen, by insisting on joining the FPOs and cooperative societies and the PMMSY Scheme. FPOs are planned to be the foundation for most government agriculture programmes because they are a platform for increasing farmers' incomes. For example, the FPO guidelines were issued in July 2020, and the Agriculture Infrastructure Funds (AIF) guidelines were published in August 2020 (which use FPOs as a base). Women farmers are less likely than men to be equipped to access this plan on their own, and as a result, they may miss out on possibilities to contribute value along the agriculture supply chain.

Farmers were unaware of the subsidies and schemes; it was strange to observe these things. The training programme covered the importance and potential of the fisheries sector, and taught the participants about the Government of India's flagship scheme, Pradhan Mantri Matsya Sampada Yojana (PMMSY) and the vision to bring about the Blue Revolution through sustainable and responsible development of the fisheries sector in India. The resource person described the estimated investment cost which was Rs. 20050 crores comprising (i) Central share of Rs. 9,407 crore, (ii) State share of Rs. 4880 crore and (iii) Beneficiary share of Rs. 5763 crore under two separate components, namely (a) Central Sector Scheme (CS) and (b) Centrally Sponsored Scheme (CSS).

Farmers were informed about the 500 Fish Farmer Producer Organisations/ Companies (FFPOs or Cs) that would be established, as announced in the Union Budget 2020, to economically empower fishers and fish farmers and strengthen their negotiating power. The remaining 200 will be established through convergence with the Department of Agriculture, Cooperation, and Farmers Welfare, Ministry of Agriculture and Farmers Welfare's continuing FPO scheme.

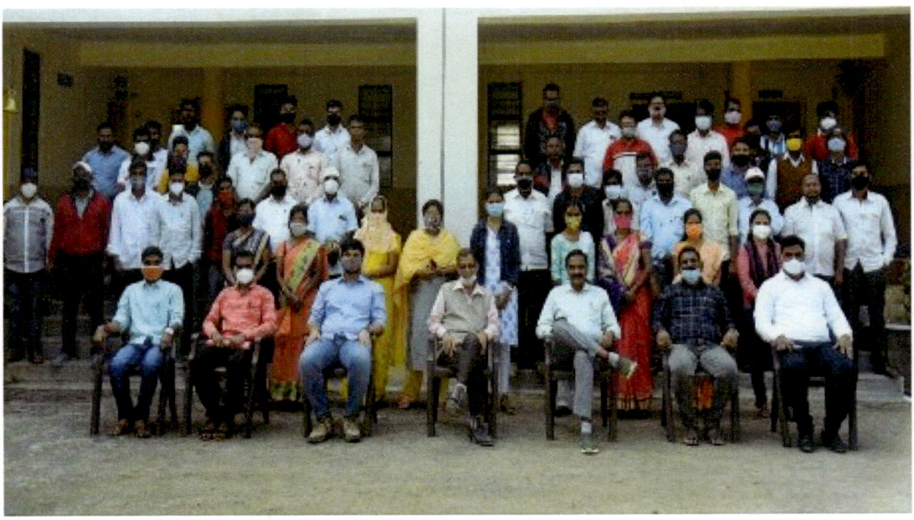

Fig. 8.4: Group photo of participants

Table 8.1: Participants participated in the training programme

Area	No. of Beneficiaries		
	Male	Female	Total
Mahankal	31	19	50

Farmers were very happy to know about FPOs and PMMSY schemes in 4 elaborative sessions,which definitely gave the formation of FPOs a boost. Awareness should be everywhere in India. Because it's high time to improve the production of fish as well as the economy of small- scale farmers, which basically depends on fish farming. The training institute was very grateful to

MANAGE, Hyderabad, for funding this training programme to educate the fisher community.

9

Integrated Farming System An Entrepreneurial Model

Sushrirekha Das and Shahaji Phand

*National Institute of Agricultural Extension Management (MANAGE)
Hyderabad, Telangana*

KVK Ramkrishna Ashram in Kolkata undertook a 5-day SCSP (Schedule Class Sub-Plan) training programme from MANAGE, Hyderabad, to develop integrated farming practices and encourage young people to choose farming as a career. West Bengal's status as the highest fish producing state in India and its leadership in fish seed production present a significant opportunity for farmers and entrepreneurs in the fishery sector. Fish is widely consumed throughout the nation, especially in West Bengal, where it is estimated that each person consumes 15.6 kg of fish per year, compared to India's average of 9.0 kg. It is regarded as a major source of essential micronutrients and the primary source of animal protein. By leveraging smart farming solutions enabled by technology, farmers can increase their productivity and income, while also meeting the demand for fish and essential micronutrients in the region. The training programme could play a vital role in empowering farmers and entrepreneurs in the state to adopt innovative and sustainable farming practices and make the most of these opportunities.

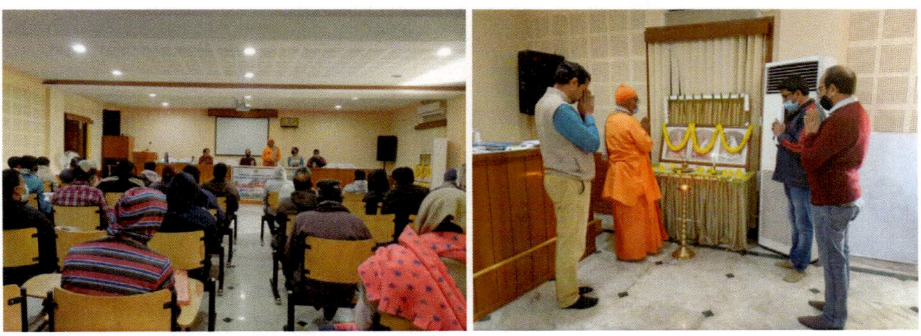

Fig. 9.1: Inaugural function at Ramkrishna Ashram Krishi Vigyan Kendra

Overview of the training programme

The programme had a total of 30 SC community participants from different districts, including 23 fishermen and 7 fisherwomen. The fact that the programme had 20 sessions, including practical classes in the field, shows that it was comprehensive and hands-on. It was also important that the training cover essential and simple techniques that can be easily adopted by farmers. The visit to integrated farms and fish farms gave the participants exposure to best practices in the industry.

Fig. 9.2: Programme coordinator explaining to the participants

Importance of training programmes

It is encouraging to see that the training programme covered both the technical and economic aspects of integrated farming for aquaculture. It is important for farmers and SHGs (Self-Help Groups) to have a thorough understanding of the system in order to make informed decisions and maximise their profits. The focus on the recirculatory aquaculture system (RAS) is especially noteworthy as it is a water-efficient method that allows for higher- density fish stocking and has the potential to increase productivity. Additionally, the mention of using GIFT (Genetically Improved Farm Tilapia) in the pond for a 120-day stocking period across three cycles per year shows the potential for fish culture to generate substantial returns. The fact that technology-backed smart farming solutions are being promoted can also encourage young people to pursue farming as a career, which is crucial for the long-term sustainability of the sector.

Fig. 9.3: Integrated farming module tour

Farmers in low-water-availability areas can use the aquaculture system to supplement their revenue, perhaps doubling their earnings. Furthermore, the sludge pumped out of the pond on a regular basis can be used to develop farm crops without the use of chemical fertilisers. Because intensive fish cultivation does not have to be limited to coastal settings, it may also be employed inland, demonstrating the system's versatility. The method, which consists of an intense fish culture pond, allows for high-density fish stocking in cages. By placing different species and sizes of fish in smaller cages in a pond, the burden may be balanced. Because this technology uses very little water, the high-density stocking of fish in various cages allows for greater flexibility in fishpond management. The varieties of fish one can culture was discussed in the training programme; Tilapia, Pangasius, Murrel, and Pearlspot, are some of the fish that can preferably be grown in the system.

Fig. 9.4: Farmers netting at the practical session

Output from the training

The beneficiaries were able to share their stories and identify the specific areas where they were facing challenges. It's was encouraging that the training programme provided opportunities for homemakers, and young people to start their own ventures and become financially independent. It's important to note that the need for awareness and training is not limited to West Bengal, but is necessary across India to promote sustainable and integrated farming practices. By sharing their success stories, these beneficiaries can inspire other farmers to adopt similar models and increase their income. It's also good to see that the institute in West Bengal acknowledged MANAGE, Hyderabad, for organising this training programme and helping to empower the rural farming community.

Fig. 9.5: Group photo of IFS training participants

All in all the beneficiaries found the 5-day training programme on integrated fish farming useful and informative. It's important to explore low-cost options like integrated farming, which not only saves time but also allows farmers to get dual benefits from the same field or area. Practising sustainable integrated farming is also important to achieve optimal productivity at a lower cost and provide a greater financial return to the rural pond farming community. The beneficiaries provided positive feedback and testimonials about the programme, which reflects of the effectiveness of the training in helping them, improve their farming practices and supplement their income.

10

Women Empowerment Through Manufacturing Sea Shell Ornaments and Showpieces

Sushrirekha Das and Shahaji Phand

National Institute of Agricultural Extension Management (MANAGE)
Hyderabad, Telangana

The aim of this training programme was to empower women from the Scheduled Caste community by providing them with skills to create handmade seashell jewellery and other handicraft products, which they could then sell to supplement their income. The training was conducted in a modular format, which included soft skill development, theoretical sessions, practical sessions, and field visits. The training programme was conducted in Keelathilayenthal Village in Ramanathapuram District, which is known for its rich marine life and abundance of seashells.

About the seashell-making-entrepreneurship model

The training programme was funded by MANAGE, Hyderabad, and was overseen by Dr. Shahaji Phand, Programme Director, and Dr. Sushrirekha Das, Coordinator. All 30 participants were women from the fishing community who were interested in learning seashell ornament- making skills. They were taught how to create various ornaments, such as necklaces, rings, bracelets, and other handicraft products, using seashells of different shapes, sizes, and shades. The fishing community considers seashell collecting a part of their fishing process, and the training programe aimed to provide additional income for the women in the villages who were already skilled in making primitive handcrafted shell products. The programme provided an opportunity for these women to develop their skills and sell their products in a wider market, which could help improve their socio-economic status.

Fig. 10.1: Session and group discussion with participants

In this case, it was especially applied to fashion accessory crafts that will increase economic value, utilisation value, and aesthetic value. The processing of shells in the area was not limited to raw shells that were reprocessed into products, but there were a number of fusion processes so that new forms were created and could be processed in accordance with the buyers. Venturing into a new business module would enhance the chances of sustainable and profitable business by women entrepreneurs. Once the training get completed, the members could form self- help groups or a cooperative society. Via this SHG, or the cooperative society, they could avail of loans with a subvention rate of interest and capital subsidy.

Fig. 10.2: Training at SIPPO

Training objectives

The trainer trained the participants on the types of ornaments made by the seashell craft method, and also said it is a very unique craft for making decorative items like

- Making beautiful wall hangings using cut shells.
- Decorate a mirror using sea shells and pearls.
- Seashell Key Chains
- Size Door Curtains (Full Curtains and Half Curtains)
- Hanging Lamps (Different Models)
- Seashell Garlands, Malai
- Seashell show case pieces.
- Seashell- fashion jewellery items made from shells, etc.

These shell crafts are more aesthetic because of their individual colour combinations and texture.

The importance of training

The invited resource person covered the sessions by teaching the procedure of ornament-making. He has explained it step by step, as mentioned in the bullet points below.

- Shells are collected from theshore directly.
- Before cleaning the shells, they are soaked in water for 10 days to remove flesh from the shell.
- After 10 days, shells are cleaned using bleaching powder. Cleaned shells are dipped in HCL for further cleaning.
- After cleaning in HCL, water washing is carried out.
- To further clean them, shells are dried in sunlight.
- Finally shell polishing is carried out using hand and machine-made brushes.
- The processing of seashells, in this case, is especially applied in fashion accessory crafts that will increase economic value, utilisation value, and aesthetic value.

Fig. 10.3: Learning by doing

Impacts

The participants found the training programme to be informative and useful. Seashell ornament making seems to be a flexible and accessible business opportunity for a wide range of people, which is a big advantage. The products are lightweight and non-perishable, as this makes transportation and storage easier. Additionally, not needing access to electricity means that artisans can work from home or in remote locations. All of these factors could help making seashell ornament a viable and sustainable business option for people in the area.

Fig. 10.4: Decor items made from sea shell

Fig. 10.5: Group photo of trainers at SIPPO

11

Sustainable Shrimp Farming for Livelihood Improvement and Income Generation

Sushrirekha Das and Shahaji Phand

National Institute of Agricultural Extension Management (MANAGE) Hyderabad, Telangana

ICAR SCAD Krishi Vigyan Kendra Thoothukudi District, Tamil Nadu, organised a 3 -day SCSP training programme on inland shrimp farming during 21st - 23rd December 2021, sponsered by MANAGE, Hyderabad. SC/ST communities are given training on sustainable shrimp/prawn culture so as to enable them to take up shrimp culture as a means of self-employment or to get employed in shrimp farms by other farmers, thereby earning a livelihood. The participants were provided with hands-on training on pond preparation, stocking, feeding, water quality management, disease control, harvesting, and post-harvest processing of shrimp. In addition to this, they were also given insights into market trends and opportunities for shrimp farming. The participants were enthusiastic and showed keen interest in learning about the latest shrimp farming techniques and practices.

Fig. 11.1: A view of the inaugural session

Objectives

To promote techno-commercial skills among the aspiring participants, the training programme sensitised the trainees about the importance of adhering to biosecurity, BMPs, and food safety standards in the successful adoption of shrimp farming in the inland. A complete package of practices for shrimp farming in inland saline areas, along with major constraints, remedial adaptive measures, and guidelines, was given to the participants through effectively tailored technical sessions, coordinated by programme coordinator Dr. T. Masanaselvam, Senior Scientist and Head, ICAR SCAD KVK, Thoothukudi, Tamil Nadu.

Fig. 11.2: Training sessions conducted by KVK, SCAD, Tamil Nadu

Importance of training

It is important for farmers to have access to technical knowledge and expertise to ensure successful and sustainable farming practices. The field visits organised by KVK SCAD allowed participants to gain practical knowledge and insights into different farming regions and practices. The opportunity to interact with other farmers and entrepreneurs from different states also provided a platform for knowledge sharing and problem-solving. The training programme not only focused on technical aspects of shrimp farming but also addressed broader issues such as biosecurity, BMPs, and food safety standards. This helps to promote sustainable and responsible farming practices while also ensuring product quality and safety. The forum provided a space for farmers to raise their concerns and challenges, which can be addressed by the government and technical experts. This highlights the importance of such training programmes in promoting inclusive and sustainable development in the agriculture sector.

Fig. 11.3: Farm visit by the trainees

On-farm demonstration

Demonstration programmes are indeed an effective way of transferring knowledge and skills to farmers. By seeing the technology in action and being able to ask questions and receive guidance from experts, farmers can learn how to implement new farming techniques more effectively. It's also great to see that the programme benefits not just the farmers who host the demonstration, but also their neighbours and other farmers who attend. This can lead to a multiplier effect, as the knowledge and skills learned in the demonstration can be spread throughout the community.

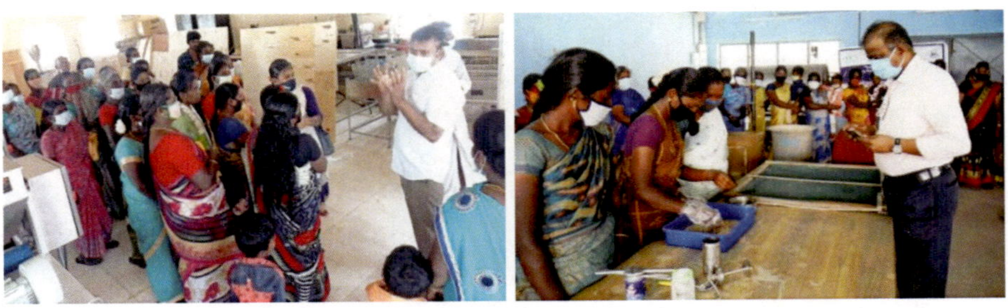

Fig. 11.4: Demonstration at a farmer's farm, Tamil Nadu

Impact

The training programme covered a wide range of technical knowledge related to shrimp farming, including important aspects such as biosecurity, water and feeding management, disease surveillance, and safe effluent disposal. The programme seems to have been designed to cater to the needs of medium-sized and marginal farmers in coastal areas, who may have traditional knowledge but lack the necessary technical skills to improve their farming practices.

The training could potentially help farmers adopt environmentally friendly shrimp farming methods, which could be a step towards sustainable and responsible aquaculture practices. It is also encouraging to note that the participants were happy to receive their training certificates, which could help them apply for schemes and subsidies to further improve their farming practices and livelihoods. Overall, the programme seems to have been well-designed and implemented, and has the potential to benefit a large number of farmers in the region.

Fig. 11.5: Group photo at the field-visit site

12

Entrepreneurship Development by Manufacturing Value- Added Fish Products

Sushrirekha Das and Shahaji Phand

National Institute of Agricultural Extension Management (MANAGE)
Hyderabad, Telangana

A one -day SCSP training program was organised by Mitraniketan Krishi Vigyan Kendra (ICAR), Kerala on 17 December 2021. The training theme was "Market opportunities for value-added Fish Products." 4 male and 46 female participants participated in the training program. The training was taken at KVK, Mitaraniketan, Kerala.

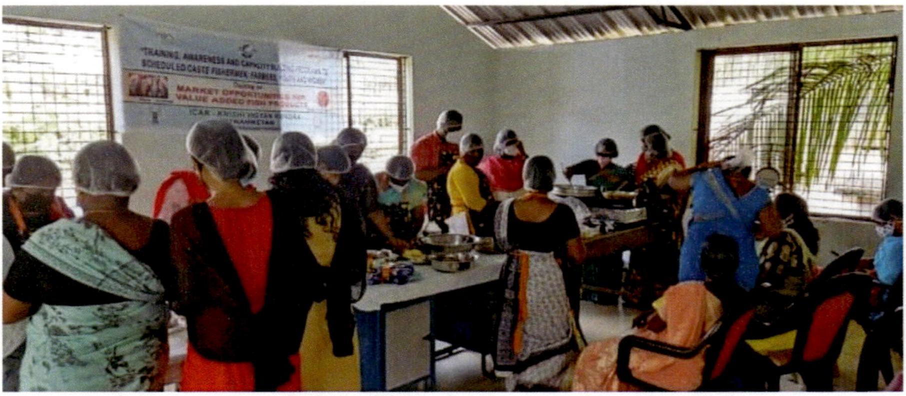

Fig. 12.1: A women group was selected for the training.

The sessions were oriented towards theory and practice. The training modules and session topics delivered on various value-added fishery products were arranged for the beneficiaries to transfer their expertise in fishery products and equip them to establish entrepreneurship in fish and fishery products and market the fishery products. In addition to providing financial support for women, the programme would also help pregnant women and adolescent girls with their nutritional needs.

Fig. 12.2: Training session taken by resource persons of KVK, Mitaranietan, Keral

In 5 days of the training programme a total of 20 sessions were held, out of which 16 were technical sessions and four sessions were field visits. Technical sessions were handed out by experts from the Dr. MGR College Ponneri faculties. The course content was hands-on training on hygienic handling of fishery products, preparation of fish steaks, fillets, fish fingers, condiment-incorporated fish products, coated fish products like cutlets, fish balls, fish burgers, fingers, cured and dried products, smoked products from freshwater fishes, preparation of butterfly shrimps, PD and PUD shrimps and coating, packaging of value-added products, waste utilisation, and marketing, etc. The participants were also oriented towards the concept of Agri-Business incubation as well as the preparation of a business plan for the establishment of a fish processing unit as a business venture, in which discussions were held regarding the equipment needed for establishing a fish processing unit, highlighting the cost-benefit analysis of the fish processing unit. Assistant Professor Dr. N. Murli lectured about the types of fish that are suited for producing various goods. Dr. N. Murli also gave a cooking demonstration for the dishes.

Fig. 12.3: Sessions taken by faculties of Dr. MGR College, Ponerri

Objectives of the training programme

Value-added seafood products are an excellent way for fisherwomen to boost their income and avoid selling their catch at a basic price. These products can be easily created by using raw fish, and the manufacturing process is relatively simple. For example, fish pickles, fish-incorporated pasta, fish-incorporated noodles, and ready-to-eat fish curry are all value-added seafood products that have the potential to generate higher profits for fisherwomen. Furthermore, younger generations are often interested in trying new and innovative products that are readily available. Therefore, by creating value-added seafood products, fisherwomen can tap into this market and build their businesses. This will not only enhance their individual potential but also contribute to the societal advancement of their communities. In conclusion, value-added seafood products offer a promising opportunity for fisherwomen to enhance their livelihoods and contribute to the economic growth of their communities.

The KVK Mitaraniketan is planning to impart special training in making value-added seafood products and their marketing to fisherwomen members of self-help groups in coastal villages. This is a wonderful initiative that can not only increase their income but also promote entrepreneurship and provide healthier and more nutritious options to consumers. The training in making value-added products such as fish pickles, prawn pickles, prawn vadai, fish vadagam, masala dry fish, prawn biriyani, fish biriyani, prawn pakoda, prawn and crab kolla balls, fish and prawn cutlet, and prawn omelette can equip these fisherwomen with new skills that can help them start their businesses. The collaboration with MANAGE, Hyderabad, can also bring more expertise and resources to this initiative, leading to a more significant impact.

The women in Tirupalaikudi village showed keen interest in becoming entrepreneurs in the fishery sector. The training provided to them is comprehensive, covering a wide range of technical areas, including fish preservation, hygiene, and the production of value-added goods like pickles and spices. The women have taken this opportunity seriously, recognising that they have the potential to create a sustainable source of income for themselves and their families. It is encouraging to note that some kiosks have already been opened in Tamil Nadu, selling fresh and dry fish, with plans to expand into other offerings such as freshly prepared fish-based meals and snacks. This demonstrates that the women are not just learning new skills, but also applying them to create innovative business ideas that can potentially generate higher margins. Women are committed to maintaining high standards of hygiene and quality in their products. This is important not only for customer satisfaction but also for building a loyal customer base.

Overall, the training programme and the women's entrepreneurial spirit offer a glimmer of hope for a sustainable future for the fishing community in Tamil Nadu. It is essential that such programmes are continued and expanded to ensure that more women have the opportunity to become self-reliant and financially independent.

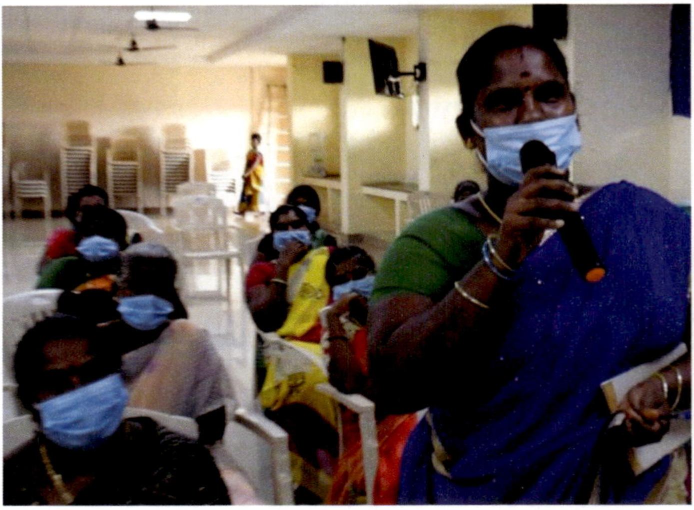

Fig. 12.4: Training feedback given by participants at Dr. MGR Fishery College Ponerri, Tamil Nadu

Table 12.1: Participants participated in the training programme

Area	No. of Beneficiaries		
	Male	Female	Total
KVK, Mitaraniketan, Kerala	04	46	50
Dr. MGR Fishery College, Ponneri, Tamil Nadu	01	29	30
Total	**05**	**75**	**80**

Aquapreneurship development

Value addition in the fishery sector involves processing, packaging, and marketing fish and fishery products in a way that enhances their value and marketability. By doing so, fisherwomen can earn higher incomes, and consumers can benefit from healthier and more nutritious food options. Additionally, value-added products can increase the shelf life of fish, reduce waste, and create more opportunities for employment and entrepreneurship in the fishery sector. MANAGE's focus on this training theme shows that there is a growing demand for value-added seafood products in the market and that there is great potential for fisherwomen to tap into this market to improve their livelihoods.

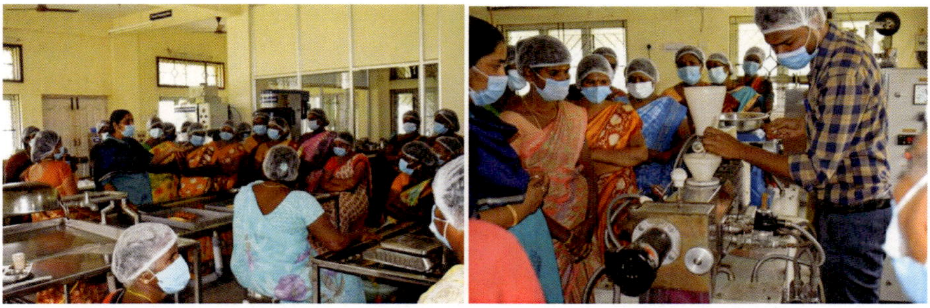

Fig. 12.5: Demonstration of the preparation of value-added fish products

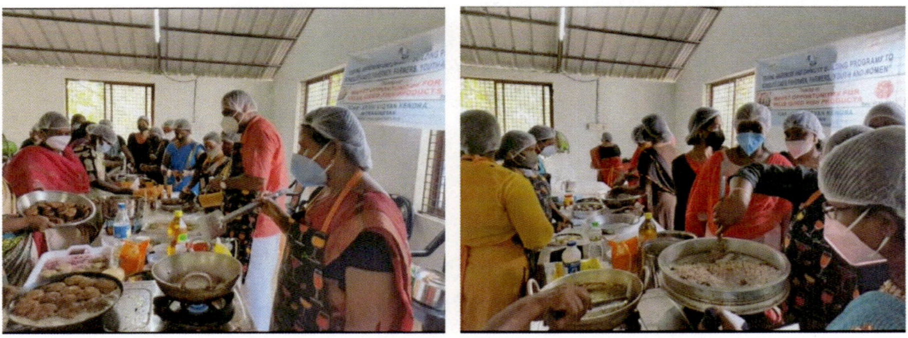

Fig. 12.6: Demo class on the preparation of fish products at KVK, Mitaraniketan, Kerala

Impact

It seems that the next steps involve providing the women with the necessary resources to turn their newly acquired skills into a successful business. They have come up with various solutions, such as catering, more shelf-stable items, versatile packing equipment, and home delivery, but they will need assistance in implementing these ideas. The training programme has taught them how to develop a business plan, project expenses and revenues, assess risks, and seek government funding opportunities. The women have already developed value-added fish food products such as fish cutlets, fish pickles, shrimp pickles, and fish, which were produced in a hygienic manner and packaged attractively. Market linkages have been established for these products through various interactions, and they have the potential to generate additional income for the women. It is reported that 30 participants from the Pulicat and Ponneri regions attended the training programme and benefited from it. The training program on "Valorization of fish and fishery products" is one of the best learning training programs for coastal fisher women community; they can run numerous women-led projects aimed at diversifying fishing households' livelihoods. The

5 days training program highlighted the entrepreneurship model, which can be a path for the coastal fisher family, which depends on the sea only to get their daily food on the plate. This is really a blessing for those trained in the training programme.

Fig. 12.7: Group photo with participants of KVK, Mitaraniketan

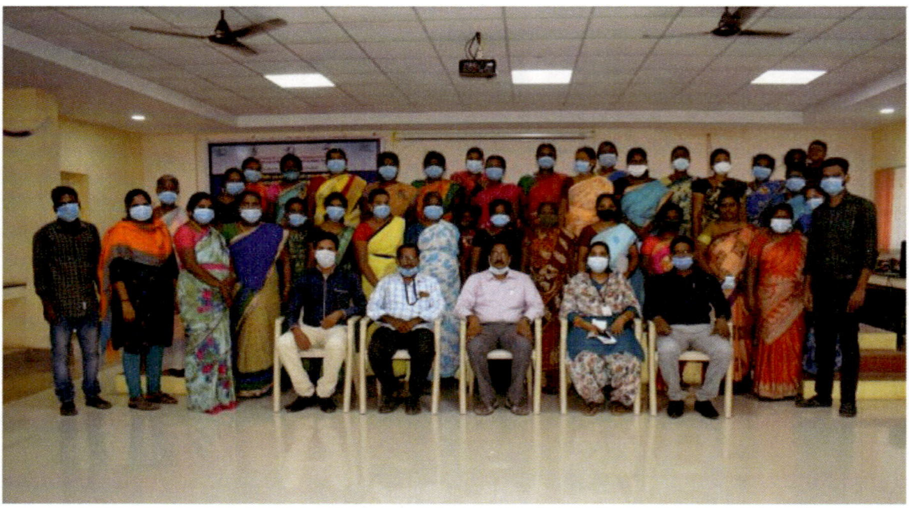

Fig. 12.8: Group photo with participants of Dr. MGR Fisheries College and Research Institute, Ponneri, Thiruvallur District

13

Seaweed Farming: From Sea to Plate

Sushrirekha Das and Shahaji Phand

National Institute of Agricultural Extension Management, (MANAGE) Hyderabad, Telangana

There are various training programmes being conducted in Tamil Nadu to promote seaweed cultivation and its various applications. The one-day SCSP training programme organised by ICAR-RVS KVK, Tiruneveli, Tamil Nadu, and the three-day SCSP training programme on seaweed for food and industrial application by Jehovah Shamma Aqua One Centre, both sponsored by MANAGE, Hyderabad, seem to have been well received by the coastal population of Tamil Nadu. The training programmes covered topics such as the status and distribution of seaweeds on the Indian coast, by-products from marine seaweeds, and nutraceutical products from marine seaweeds. These are important areas of knowledge that can help the coastal population explore the various opportunities available in seaweed cultivation and to utilise the seaweed species to their full potential.

Fig. 13.1: Training at RVS-KVK

Fig. 13.2: Training at Jehovah Shamma Aqua One Centre, Tamil Nadu

Objectives of training the programme

The main objective was to train the coastal fisher women, men, youth, and self-help groups in cultivating seaweed, employing the appropriate technology, increasing biomass productivity of seaweeds for industrial requirements, and encouraging entrepreneurship development on seaweed-based activities among community-based organisations by KVK-RVS and Jehovah Shamma Aqua One Centre, Tamil Nadu.

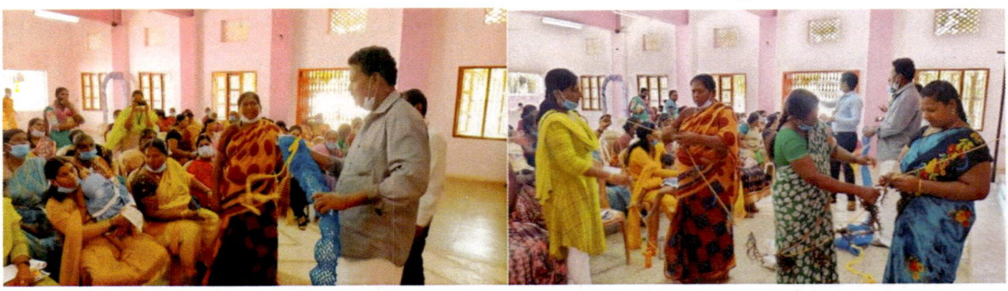

Fig. 13.3: Training demonstration at RVS-KVK, Tamil Nadu

Fig. 13.4: Training demonstration by Jehov Shaama, Aqua One Centre, Tamil Nadu

Interventions

Experts from different national institutes and industries are coming together to support seaweed cultivation and business in Tamil Nadu. The 'Make in India' scheme, which provides beneficiaries with free rafts, weed, and technology, can help them earn additional income while protecting the marine ecosystem. The support from agar companies and government agencies like the Fisheries Department and GOMBRT is also encouraging. Overall, the training programme and inputs can promote entrepreneurship and improve the socio-economic status of people in the Asia-Pacific region. In the 'Make in India' scheme, each beneficiary would be given five rafts, weed, and the technology free of charge under the concept, which aims to transfer the benefits of technology developed at research stations to the general public. This would help them earn additional income and prevent them from harvesting seaweed indiscriminately in protected areas.

Benefits of training programmes

It's great to hear that the training programmes generated much excitement and interest among the beneficiaries in seaweed cultivation. Seaweed cultivation has a lot of potential as a source of income for coastal communities, and it's encouraging to see efforts being made to promote entrepreneurship in this area.

Seaweed cultivation is very important for the Asia Pacific region, according to the expert members, because seaweed output has not increased as projected in this region. They went on to say that seaweed has a lot of potential as a food and as an industrial input, with applications ranging from cosmetics to pharmaceuticals. There are 720 species of seaweed in India, with 60 of them being commercially important. They can be found in large numbers around the shores of Tamil Nadu, Gujarat, Lakshadweep, and the Andaman and Nicobar Islands. Trainers explained that agar is used as a gelling and thickening ingredient in the confectionery and baking industries, as well as a stabiliser in the preparation of cheese and salad dressings. It is also used as a liquor clarifying agent. It is used as a laxative in the pharmaceutical sector to treat constipation. It's also employed in the cosmetic and textile sectors. Knowing the facts of seaweed cultivation, trainers were very excited to start their cultivation. This highlights the importance of sustainable seaweed cultivation, which can not only provide economic benefits but also contribute to the development of various industries.

Fig. 13.5: Classroom lecture to explain the processes

Impact

In one day, the training programme organised by KVK-RVS, Tamil Nadu, gave scope and opportunities in seaweed cultivation, while in three days, the training programme also assisted in the identification of bottle necks at various levels and their solutions. The beneficiaries who came for training Seaweed farmers, entrepreneurs, youth, women, and vulnerable communities were among the training's main benefactors. Both the training programmes in Tamil Nadu aimed to raise awareness of seaweed farming from a business standpoint, as well as develop an actionable, time-bound plan that would help India achieve its vision of "Atmanirbhar Bharat." The training's goal was to build a commercial seaweed farming concept through a network of cooperatives.

Table 13.1: Details of participation in different trainings

Training covered Area	Types of training programmes	Organisation	Participated Beneficiaries		
			Male	**Female**	**Total**
Tiruneveli	One Day	ICAR-KVK-RVS	05	50	50
Ramanathapuram	Three Days	Jehov Shamma Aqua One Centre	32	18	50
Total			**37**	**68**	**100**

Seaweed has a global market worth more than USD 12 billion. China, Indonesia, and South Korea are the top producers of seaweed, while India has a negligible representation at the current time, which is why all stakeholders must work together to alter this situation. Around 5000 metric tonnes of farmed seaweed were produced in India, while 25000 metric tonnes of wild seaweed were gathered. In India, the entire market value of the seaweed industry is estimated to be between Rs. 300 and Rs. 500 crore, said the presenter. Seaweed can be used as human food, animal feed/supplements, bio fuel, bio fertilizer/ growth stimulants, and for industrial uses such as nutraceuticals, cosmetics, industrial chemicals, and so on. It can also be used as a nutrient scrubber, a carbon dioxide scrubber, an oxygenator, and a nursery ground for fishermen. About USD 5 billion is contributed by seaweed food products for human use.

Fig. 13.6: Farmers after training stated seaweed cultivation

There were 100 beneficiaries who were trained in both training programme, majority of when were women, and the breadwinners of their families. They are now profiting handsomely from seaweed farming. Several women's organisations have benefited from MANAGE, Hyderabad, by initiating training, and technology has ensured the local population's livelihood while also empowering them.

14

Entrepreneurial Opportunities in the Fisheries Sector of the North East Hill Region of India

Sushrirekha Das and Shahaji Phand

National Institute of Agricultural Extension Management (MANAGE). Hyderabad, Telangana

The SCSP training programme on biofloc-based fish farming in Manipur aimed to promote entrepreneurship in fish farming by providing training on modern and innovative techniques. The training programme was sponsored by MANAGE, Hyderabad, and was conducted by two organisations in Manipur: AOC Manipur and Green Biotech Ecosolutions Pvt. Ltd. The programme was held over a period of three days at Khurkhul Village, Imphal West, Manipur. The programme focused on biofloc-based fish farming, which is an innovative and sustainable method of fish farming that uses microbial flocs as a source of nutrition for fish. The training covered various aspects of biofloc-based fish farming, including the principles of biofloc technology, system design, water quality management, nutrition management, and disease management.

The programme aimed to provide participants with the skills and knowledge necessary to start and operate a successful biofloc-based fish farming business. It also aimed to promote entrepreneurship in the fish farming sector in the North East region of India, where there is a growing demand for fish due to the increase in human population. By providing training on modern and innovative techniques like biofloc-based fish farming, the programme aimed to encourage farmers in the region to adopt a more entrepreneurial approach to aquaculture. This, in turn, would help to increase the productivity and profitability of fish farming operations in the region and meet the growing demand for fish.

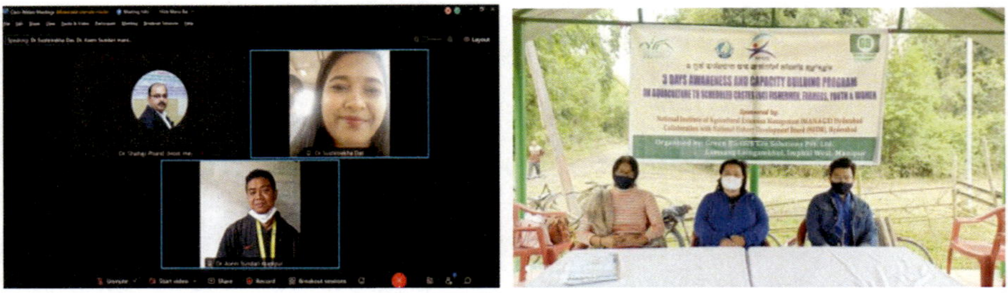

Fig. 14.1: Awareness and capacity building training programme at Manipur

Fig. 14.2: Awareness and capacity building training at West Manipur

Objectives of the training programme

a) To attract enhanced private investment in fisheries and aquaculture sector.

b) To enhance production, productivity, and profitability across the value chain by achieving economies of scale, encouraging technology uptake, and addressing value chain gaps.

c) To foster linkages among producers, aggregators, processors, and exporters for better price realisation and enhanced incomes.

d) To generate sustainable employment and livelihood opportunities, especially among youth and women in rural and semi-urban areas.

e) To leverage the existing knowledge capital and expand into newer, untapped markets.

f) To create an ecosystem for the growth of entrepreneurship in the fisheries and aquaculture sector.

Fig. 14.3: Training at Khurkhul Village, Imphal West, Manipur

Training background

Assamese farmers typically raise cattle on their homesteads. They were questioned about whether they felt that training in integrated livestock fish farming was crucial for them because feed is the primary input in fish aquaculture and can be substituted by recycling livestock waste. The majority of farmers (88%) said that such training is not necessary for them. The majority of farmers (80%) chose water quality management training because they were aware of how water quality affects fish production. However, only 80% of farmers were concerned about the diet and feed of their fish. Only 10% of farmers thought it was absolutely necessary. The main farming activity in the state is agriculture, particularly rice farming. Aquaculture needs to become more intensive in order to keep up with the growing demand, better combat poverty and malnutrition, and improve its ecological sustainability. Sustainable aquaculture will be able to take the place of conventional aquaculture thanks to new technologies. This increase in aquaculture production must occur in a sustainable manner through the use of innovative farming interventions, including integrated farming, aquaponics, and recirculatory aquaculture systems (RAS), in order to enhance the socioeconomic situation of the farmers (RAS) and biofloc-based farming.

Fig. 14.4: Participants at the integrated farm, Ecosolutions Pvt.Ltd.

Scope for aquaprenurship development

It is quite encouraging to see that the fishing industry is being viewed as a means to foster entrepreneurship and promote rural development in African and Asian countries. The training programme described seems comprehensive and covers a range of important topics related to fishery entrepreneurship, including post-production management and policy implementation.

The focus on ornamental fisheries is interesting, and it is encouraging to hear that the NFDB envisions a cluster approach to the holistic development of this industry. Community-based ornamental aquaculture can provide benefits beyond income generation, such as conservation benefits and food security. It is important to identify and address the challenges and roadblocks that this project may face to ensure its success. Overall, the training programme seems like a positive step towards promoting entrepreneurship and job creation in the fishing industry in India.

The goal of both the training programmes was to bring together a diverse group of stakeholders to discuss ways to promote entrepreneurship and job creation in the fishing industry. Furthermore, the training programme aimed to identify and resolve bottlenecks at the policy and implementation levels. The Indian fisheries sector, which earns a lot of foreign exchange, encourages the expansion of related sectors, ensuring the supply of affordable, nutritious food for socioeconomically disadvantaged small farmers. As a result, there is a critical need to generate a trained workforce for fishing entrepreneurship.

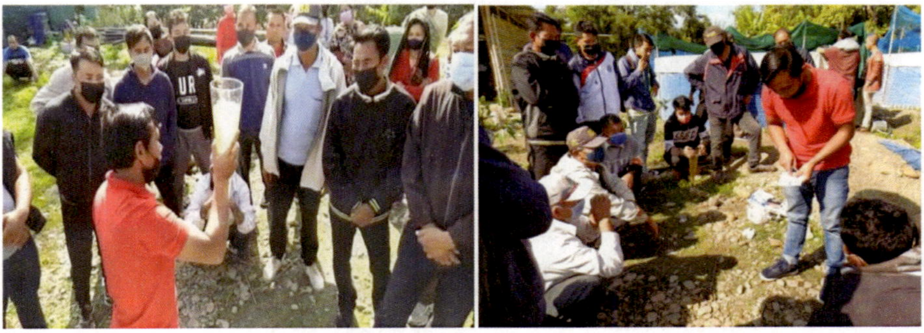

Fig. 14.5 Demonstration about biofloc fish farming

Challenges in aquapreneurship development

In Assam, fishing has always been a caste-based industry. Fish were primarily found in flood plain lakes, which were originally uncontrolled natural water basins. Fisheries activity has historically been carried out by three main categories of people: (1) individuals who catch fish for their own daily food; (2) members of the fisher community; and (3) rural fisher entrepreneurs (leaseholders). For food, the average person used to catch fish every day, whereas fishermen worked full-time. Aquaculture is a very young area of food production, and Assam is constantly changing it.

Fig. 14.6: Practical demonstration of FCO preparation to the participants

Fig. 14.7: Group photo by Ecosolutions Pvt. Ltd.

Fig. 14.8: Group photo by Aqua One, Manipur

Impacts

There is a potential for aquaculture to be transformed from a poverty relief programme to a respectable income-generating enterprise in Assam. Providing extensive training to farmers on the most recent technological breakthroughs in aquaculture is an important step towards achieving this goal. Awareness camps on recent advancements in aquaculture technologies can be an effective way to introduce farmers to new methods and techniques. It is also encouraging to see that various stakeholders, including scientists, government officials, entrepreneurs, and training participants, were actively involved in the training programmes and contributed to their success.

It is important to continue providing support and training to farmers to help them improve their productivity and profitability in aquaculture. By doing so, it is possible to achieve sustainable and profitable aquaculture systems that benefit both farmers and the wider community.

15

Empowering Rural Youths Through Ornamental Fish Breeding and Farming

Archana Sinha

ICAR-Central Inland Fisheries Research Institute (ICAR-CIFRI)
Barrackpore, Kolkata, West Bengal

Ornamental fish trade is a part of the culture of fisheries in India and is an important sector for livelihood and income generation, especially for women. Recent trends in production and demand share assume that during the next five years, domestic trade will double from the present. The growth of any business depends on technology, management, market, and infrastructure. The ornamental fish industry includes breeding of exotic live-bearer and egg layer fish, captive breeding of indigenous ornamental fish, the development of small-scale enterprises, backyard fish breeding and culture activities in rural areas, and ancillary units with high potential for large- scale exports as part of the multibillion-dollar global trade of the ornamental fish sector. Women are sensible and caring enough to take care of and nurture the small, tiny, colourful fish in their backyard units of ornamental fish. One of the important sectors of the ornamental fish industry is aquarium making and decoration, which is not only women friendly but also ensures earnings. The training provided opportunity, involvement, income generation, and technology support to women through ornamental fish breeding and culture for their empowerment.

Objectives

To impart training in ornamental fish breeding and culture to educated unemployed women to start their own business.

Interventions

The ICAR-Central Institute of Fisheries Education's put efforts to train people in Betiah, Bihar, in ornamental fish breeding and culture. The six-day training programme seems to have been comprehensive, covering various aspects of ornamental fish breeding and culture, including export potential, marketing methods, and bank finances available in this sector. The practical classes and demonstrations on the construction and maintenance of an aquarium, identification of different species, water quality management, feeding strategies, and packaging of ornamental fish would have been very beneficial for the trainees. The visit to developing ornamental fish farms in Aamtala and Howrah of West Bengal and the group discussion with the Managing Director of Malabar Tropicals, a leading export company in ornamental fishes, would have provided valuable insights into the norms and procedures for the export license and trend in the export market.

It's encouraging to see such initiatives being taken to promote entrepreneurship and job creation in the fishing sector. Such training programmes could potentially help generate employment and income for people in rural areas, and contribute to the socio-economic development of the region.

Fig. 15.1: Glimpses of training the programme arranged at ICAR-CIFE, Kolkata Centre

Interventions

Women trainees were acquainted with the principle of breeding ornamental fish. Opportunities were given to them to handle the hatchery of ornamental fish. They bred fish with their own hands.

Fig. 15.2: Lecture delivered by the organizers

Fig. 15.3: Trainees observing the breeding process

Fig. 15.4: Participants busy in training and learning activities

An exhibition was organised in a seminar held at Betiah, Bihar. The special attraction was a competition for aquarium making and decoration by the beneficiaries. About 65 women trained at ICAR-CIFE, Kolkata Centre, participated and displayed the aquarium with innovative ideas. Ms. Priti Gupta and Ms. Shahin Nazar received the first prize from Shri Dilip Kumar, District Magistrate, West Champaran. The initiative taken by Shri Dilip Kumar, IAS, District Magistrate, West Champaran, to conduct such a seminar and training at Bettiah was highly appreciated. It was one of its kind, with the active participation of fish farmers, entrepreneurs, scientists, GO's, and NGO's brought together on one platform.

16

Ornamental Fish Breeding and Farming: A Source of Livelihood for Rural Women of West Bengal

Archana Sinha

ICAR-Central Inland Fisheries Research Institute (ICAR-CIFRI)
Barrackpore, Kolkata, West Bengal

Ornamental fish breeding and culture: an innovative scheme for the development of rural women." Funded by the Department of Biotechnology, Ministry of Science & Technology Govt. of India, New Delhi

Fig. 16.1: Ornamental fish breeding facility

Selection of members and group formation

The location was selected to impart the training on ornamental fish culture, based on the PRA, group discussions, and meetings with NGOs, village panchayat authorities, and members of the area. Raichak, Kamar Pool, and Diamond Harbour Road were the working locations. Groups formed by identifying the interested women from the selected location. In each group, there are 25 women registered for the training. The selected members were introduced to the ornamental fish farms located at Amtala to appraise the working possibilities and viability of the activities.

Fig. 16.2 Women participants in the training session

Fig. 16.3: Kamarpool (Sarisa Ashram), Raichak, Diamond Harbour Road, Dist.24 Parganas (S), West Bengal.

(A) Construction of infrastructure

Construction of infrastructure for demonstration and training of ornamental fish breeding to the women group completed on site.

Fig. 16.4: Cement tanks are constructed on site for hands-on training.

(B) Backyard hatchery and nursery for the culture of ornamental fish

Fig. 16.5: Participants in the backyard hatchery and nursery

The ornamental fish breeders in Kolkata supply young ornamental fish to different retailers who sell aquaria and aquaria accessories. The success story of SWARNALI Women's Cooperative Society for ornamental fishes in South 24 Parganas of West Bengal encourage women from other areas to adopt this innovative scheme of fisheries.

Training could improve the livelihood of 30 underprivileged rural women by successfully completing a project on backyard ornamental fish breeding and management in the village. The net income is in the range of Rs. 20,000 to Rs. 50,000 per annum. Still, there is a wide gap between demand and supply. With the high pace of demand, the scope for development of the ornamental fish industry has also increased. Women have shown their interest and capability in adopting this trade in the areas like Howrah, the North 24 Parganas, and the South 24 Parganas of West Bengal. This may replicate for other members in the target area and by imparting training, the skill of the participants can be improved. This will definitely help the target population improve their livelihood by using aquaculture as an income generating activity.

Fig. 16.6: Visitors at the appreciated the work done by the women group.

17

Empowerment of Tribal Population Through Awareness and Training Programmes on Sustainable Fisheries Based Interventions

Madhuri Sharma, A.P. Singh, and Hari R.

College of Fishery Science, Nanaji Deshmukh Veterinary Sciences University, Jabalpur, Madhya Pradesh

Madhya Pradesh has the highest population of Schedule tribes in India. With a population of 15,316,784, they occupy 21.1% of the state's total population and 14.7% of the total tribal population of India. About 61% of the rural tribal population in Madhya Pradesh is living below the poverty line and is primarily relying on agriculture and allied industries for their livelihood (Planning Commission Report, 2010). But since the majority of them rely on unscientific subsistence farming, the income earned from such practices is very low.

Apart from the higher poverty level, the nutrition levels in Madhya Pradesh also signify a grave situation. The state has one of the highest levels of anaemic children and women in India, and this incidence is highest in tribal regions of the state. The tribal regions of Madhya Pradesh have one of the highest percentages of malnourished children in India (UNICEF, 2019). The undernutrition and anaemia in tribes are primarily due to the lack of access to quality food.

Fisheries-based income generation activities have the ability to provide a sustainable source of income and nutrient rich food to the tribes. Since the majority of the tribes in Madhya Pradesh have very little land holding and are categorised as marginal farmers (Census, 2011), fisheries can provide them with a regular source of livelihood. Fish acts as one of the best sources of protein and vitamin A.

In view of the above fact, we organised some awareness and training programmes based on fisheries, like the rearing of fish seed, the preparation of different fish products, etc., to improve the livelihood and economic level of the tribals of Madhya Pradesh.

Awareness and training programmes were conducted for the empowerment of rural tribal communities through sustainable fisheries and utilisation of water bodies for livelihood security in Gughari, Magardha, Lawar, and Gopalpur Villages of Bijadandi Block, Mandla, Madhya Pradech (M.P.).

Challenges and opportunities

The Gughari, Magardha, Lawar, and Gopalpur villages are located in Bijadandi Block of Mandla district of M.P. The tribal people living here who have a low level of literacy and are economically backward, but they are very close to nature.The people of these villages were dependent mainly on agriculture for their livelihood.

Interventions

Scientific interventions, *viz,.* natural resource identification and inventory, integrated fish farming, application of vermicompost; composite fish farming, etc.

Demonstration and implementation of process

The members of the tribal community (both male and female) were selected as the target group. As the target population was below the poverty line, alivelihood support component was also added to the programme.

Awareness Programmes

Fig. 17.1: Discussions with fish farmers in Gughari Village of Bijadandi Block, Mandla

Fig. 17.2: Distribution of Pamphlets to Farmers of Gughari Village of Bijadandi Block, Mandla

Fig. 17.3: Discussion with Fish Farmers of Gopalpur Village of Bijadandi Block, Mandla

Fig. 17.4: Discussion with fish farmers in Lawar Village of Bijadandi Block, Mandla

Hand -on training programme for the preparation of fish cutlets and fish pickle

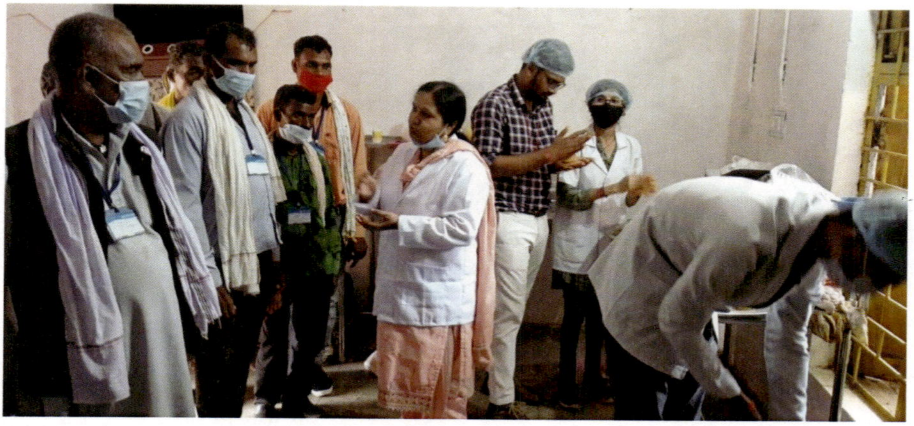

Fig. 17.5: Hands-on training for the preparation of fish cutlets for farmers in Gopalpur and Gughari villages of Bijadandi Block, Mandla

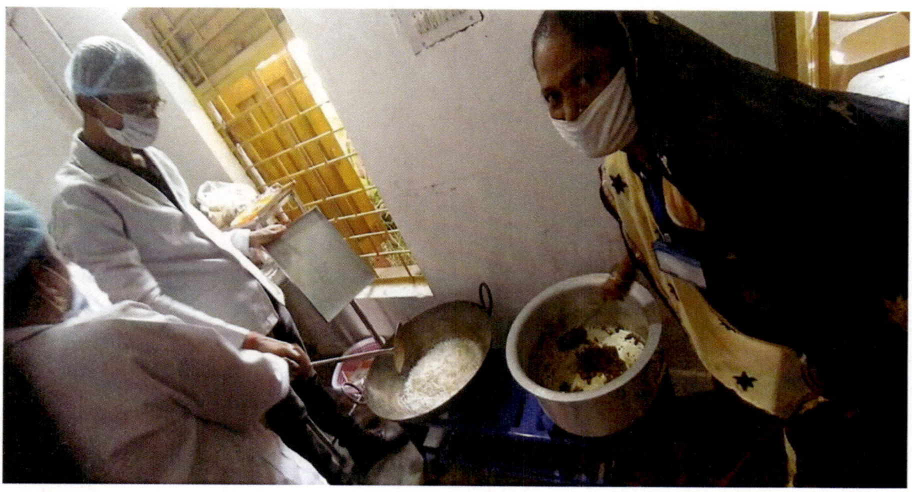

Fig. 17.6: Hands-on training for the preparation of fish pickles for farmers in the villages of Bijadandi Block, Mandla

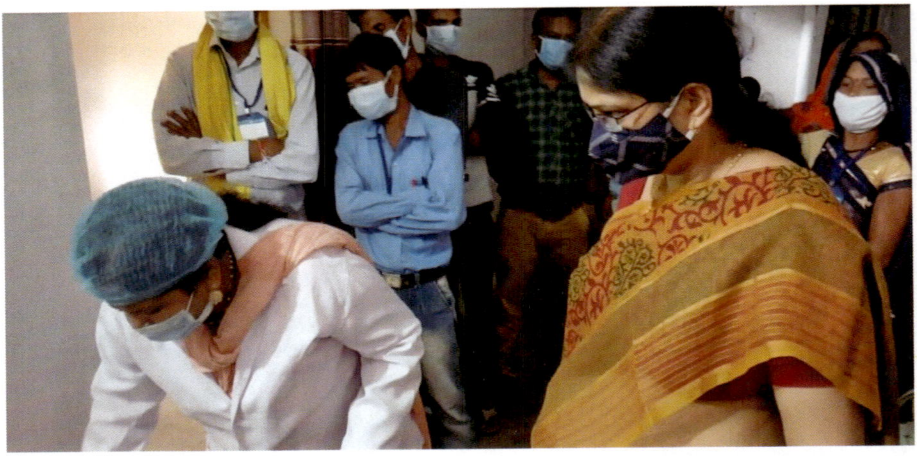

Fig. 17.7: Hands-on training for the preparation of fish pickles for the farmers in the villages of Bijadandi Block, Mandla

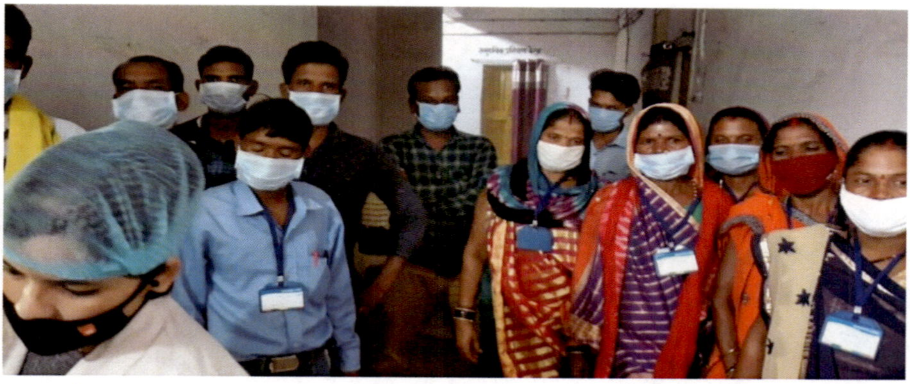

Fig. 17.8: Farmers are going through hands-on training.

Fig. 17.9: Students (BFSc final year) with fish products (fish pickles and fish cutlets)

Input and training support

Awareness and hands on training programmes were conducted on different aspects such as water quality management in freshwater aquaculture, the importance of fish in our lives, the preparation of fish pickles, the preparation of fish cutlets, the identification of commercially important fin fishes, fish disease management, integrated fish farming, composite fish farming, feed management, pre- and post- stocking management for fish culture, etc. These programme were conducted for capacity building and improvement of the social status of tribal people by the College of Fishery Science, NDVSU, Jabalpur, under ICAR-TSP programme. We distributed some input, such as fish seed fry and fingerling, fish feed, nets, lime, medicine, etc.

Fish seeds and fish feed distribution programme

Fig. 17.10: Dr. S.P. Tiwari,Vice-Chancellor, NDVSU, Jabalpur, is distributing fish feed to tribal farmers of Gopalpur and Gughari villages of Bijadandi Block, Mandla

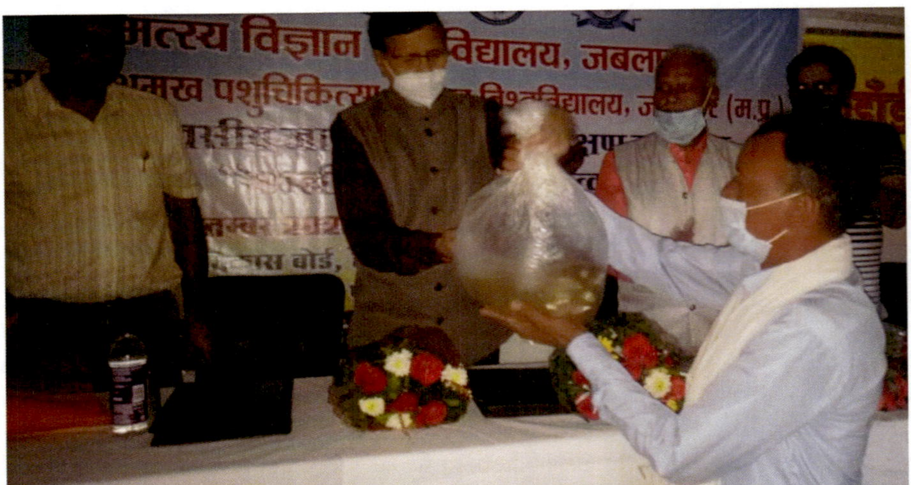

Fig. 17.11: Dr. S.P. Tiwari, Vice-Chancellor, NDVSU, Jabalpur, distributing fish seed to tribal farmers of Gopalpur and Gughari villages of Bijadandi Block, Mandla

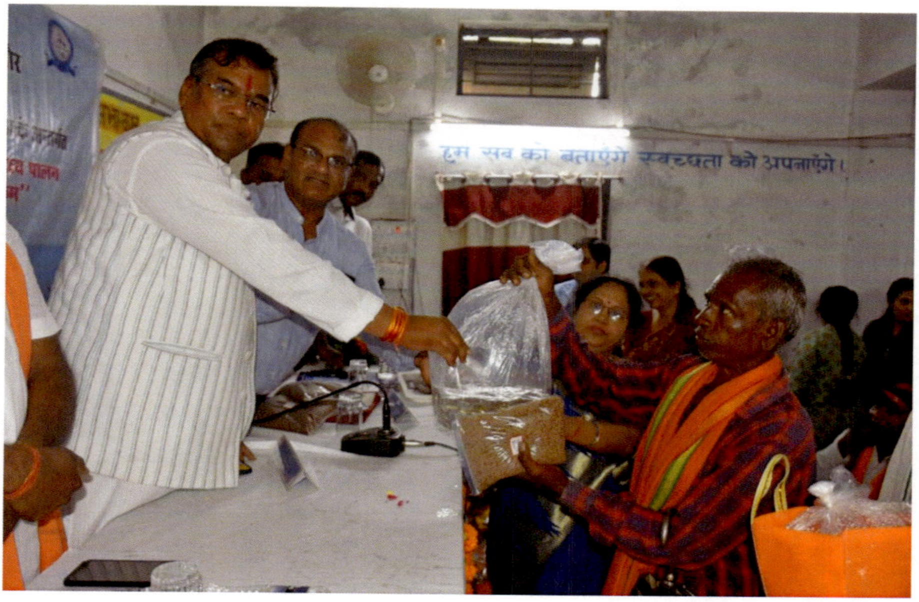

Fig. 17.12: Shri Faggan Singh Kulaste, Minister of State for Rural Development and Steel, Govt. of India, is distributing fish seed and fish feed to tribal farmers of Gopalpur, Gughari, and Magardha villages of Bijadandi Block, Mandla.

Outcomes

The knowledge of villagers on efficient water management and livelihood improvement through maximum utilisation of water bodies for fish culture was enriched through interaction and demonstration. The growth rate of stocked fish is being monitored and found satisfactory.

Impacts

Villagers and Panchayat members are now interested in adopting fisheries in available water bodies lying fallow. Women of the village are showing interest in fish culture for more income and livelihood. Almost every house is eating fish and started earning by selling fish. The rearing of major carp has added additional income and nutritional security to the poor tribal families.

18

Awareness Programmes on Pradhan Mantri Matsya Sampada Yojana and Scope for Fish Farmers

Madhuri Sharma[1] and Bimal Prasanna Mohanty[2]

[1]College of Fishery Science, Nanaji Deshmukh Veterinary Sciences University, Jabalpur, Madhya Pradesh
[2]Fisheries Science Division, Indian Council of Agricultural Research (ICAR), New Delhi, Delhi, India

Introduction

In May 2020, The Government of India launched the flagship scheme called Pradhan Mantri Matsya Sampada Yojana (PMMSY), with an estimated investment of Rs. 20050 crores, comprising a Central share of Rs. 9407.00 crores, a state share of Rs 4880.00 crores, and a Beneficiary contribution of Rs. 5763.00 crores for its implementation during a period of 5 years from FY 2020-21 to FY 2024-25 in all States/Union Territories (Guidelines on Entrepreneurship Models in Fisheries and Aquaculture under PMMSY, 2020).

The PMMSY will address critical gaps in fish production and productivity, quality, technology, traceability, post-harvest infrastructure, and management. It will further modernise and strengthen the value chain by establishing a robust fisheries management framework and promoting fishers' welfare. PMMSY also aims to empower youth and women through capacity building, innovation, and entrepreneurship (Guidelines on Entrepreneurship Models in Fisheries and Aquaculture under PMMSY, 2020).

Aims of PMMSY

By 2024-25, PMMSY envisages a target of 70 lakh tonnes of additional fish production, an average aquaculture productivity of 5 tonnes/hectare, 1 lakh crores in exports, the creation of critical fisheries infrastructure, reduction of post-harvest losses, modernising and strengthening the value chain, enhancing domestic fish consumption to 12 kg per capita, attracting private investments in the fisheries sector, and generating about 55 lakh (Guidelines on Entrepreneurship Models in Fisheries and Aquaculture under PMMSY, 2020).

Awareness programmes on PMMSY

The College of Fishery Science, NDVSU, Jabalpur, organised more than 15 awareness and training programmes on PMMSY for fish farmers, fishers, departmental officers, students, and faculty to create awareness and maximum utilisation and brood circulation of PMMSY.

Awareness and training programmes organised under the PMMSY

1. **National Training on Machhali Palan Aay Ka Sadhan:** This training organised during the COVID pandemic period from 28-30 November, 2020, by online mode for fish farmers and entrepreneurs in Hindi. Dr. B.P. Mohanty, ADG, Inland Fisheries, ICAR, New Delhi, delivered the inaugural address and special lecture on PMMSY. Experts informed in detail about the aim, criteria for apply, and different components of the Pradhan Mantri Matsya Sampda Yojna. The session was very integrative, and experts gave answers to questions from farmers about this Yojna. The training covered most aspects of management practices for fish farming. A total of eight lectures were delivered in three days. More than 80 participants from eight states of India registered and participated in this programme.
2. **Three-day National Training on "Better Management Practices for Fish Farming for Increasing the Income of Fish Farmers"-**This was also organised during the COVID pandemic period from 8-10 March, 2021 in online mode for fish farmers, entrepreneurs, and students in Hindi. Prof. R.P.S. Baghel, Dean, College of Fishery Science and Dean Faculty, NDVSU, Jabalpur delivered the welcome address. Dr. B.P. Mohanty, ADG, Inland Fisheries, ICAR, New Delhi, delivered the inaugural address. More than 56 participants' from 10 states in India registered and participated in this programme.

The training covered most aspects of management practices for fish farming. A total of nine lectures were delivered in three days, including Pradhan Mantri Matsya Sampda Yojna, Pre and Post Stock Management in Fish Farming, Feeding Management for Better Growth of Fishes, Some commercially Important Fresh Water Species of Fishes, Fish Diseases and Their Treatment, Water and Soil Quality Management for Fish Farming, Modern Techniques of Fish Farming, Fish seed and Their Transportation and Schemes of the Madhya Pradesh Govt. for Fish Farmers. These important topics were discussed during

3 days period of online training by eminent speakers Dr. B. P. Mohanty, ADG (Inland Fisheries), ICAR, New Delhi, Dr. R. S. Chouhan, Dean, College of Fisheries, G.B. Pant University, Pantnagar, Dr. K. N. Mohanta, Principal Scientist, ICAR-CIFA, Bhubaneswar, Dr. Piyush Punia, Principal Scientist, ICAR-IIFSR, Meerut, U.P.; Dr. K. Pani Prasad, Principal Scientist, CIFE, Mumbai, Dr. Ashutosh Mishra, Associate Professor, CoF, GBPUAT, Pantnagar, Dr. Sunil Kumar Nayak, Scientist & Officer In-Charge ICAR-CIFE Powarkheda Centre, Dist. Hoshangabad, Madhya Pradesh, and Shri K.L. Maravi, In-Charge Joint Director Fisheries, Jabalpur Division, Jabalpur.

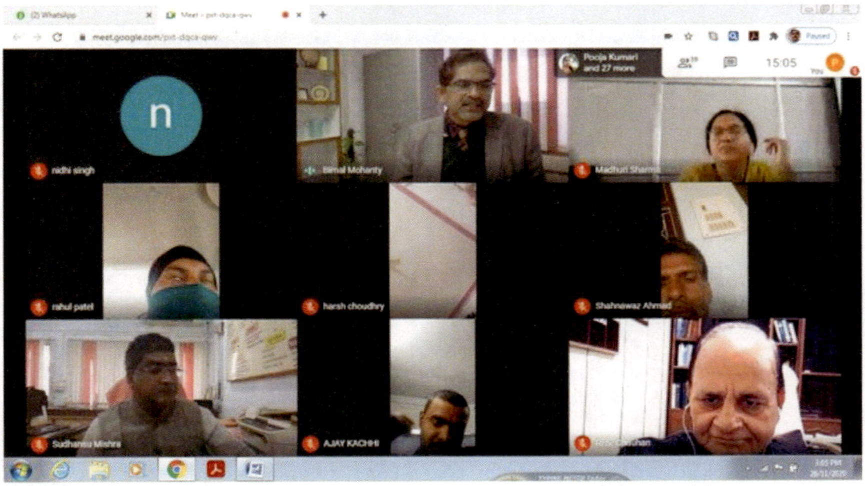

Fig. 18.1: Dr. B.P. Mohanty, ADG (Inland Fisheries), ICAR, New Delhi, delivered inaugural address on 28 November 2020. Other Speakers/Delegates present were Dr. RPS Baghel, Dr. R.S. Chouhan, Dr. S.S. Mishra, Dr. Madhuri Sharma, Organising Secretary of the programme and farmers.

Fig. 18.2: Dr. B.P. Mohanty, ADG, Inland Fisheries, ICAR, New Delhi, delivered a talk on PMMSY.

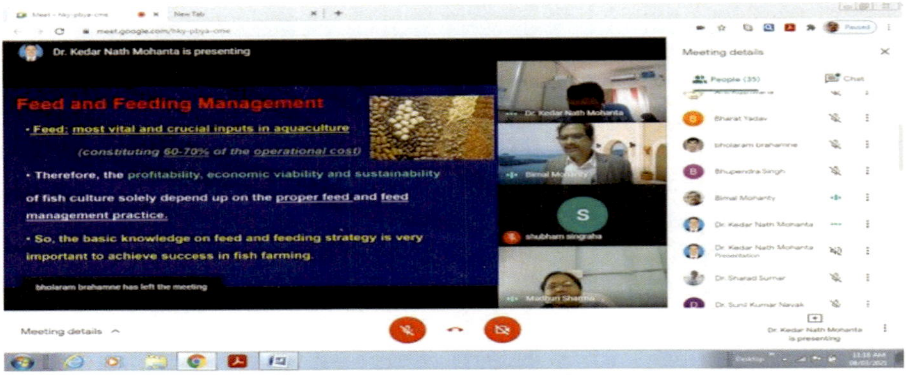

Fig. 18.3: Dr. B.P. Mohanty, ADG, Inland Fisheries, ICAR, New Delhi, delivered inaugural address on 8 March 2021. The other Guests/Speakers present were Dr. K. N. Mohanta, Principal Scientist, ICAR-CIFA, Dr. Madhuri Sharma, Organising Secretary of the programme and farmers.

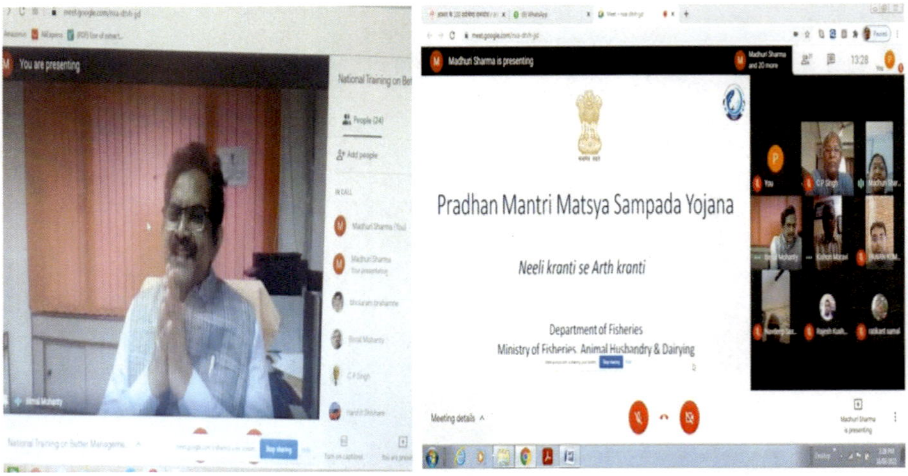

Fig. 18.4: Dr. B.P. Mohanty ADG, Inland Fisheries, ICAR, New Delhi delivering a talk on Pradhan Mantri Matsya Sampada Yojana (PMMSY)

3. Online Training andAwareness Programme on "Biofloc Culture" - This programme was organised on June 19, 2021, for students, entrepreneurs, farmers, fishermen and others. The chief guest of the program, Dr. B.P. Mohanty, ADG, Inland Fisheries, ICAR, New Delhi, delivered the Special Address in the inaugural session. The guest of honour, Shri Bharat Singh, Director, Department of Fisheries, MP, stated that the State Fisheries Department government is working on this new technique. Dr. MD Aklakur, Scientist and Officer in-Charge of ICAR- CIFE Reg. Centre, Motipur, Bihar, delivered a talk on the introduction and importance of biofloc technology. Dr. Akshaya Panigrahi, Principal Scientist, Fish and Fisheries Science, ICAR-CIBA, Chennai, delivered a talk on the components of biofloc technology. Dr. Ambika Prasad Nayak, Scientist (Fishery Science), Sakshigopal, KVK-PURI, OUAT, Odisha, delivered a talk on the application of biofloc technology for the development of sustainable aquaculture. 571 participants registered for this programme from different states of India.

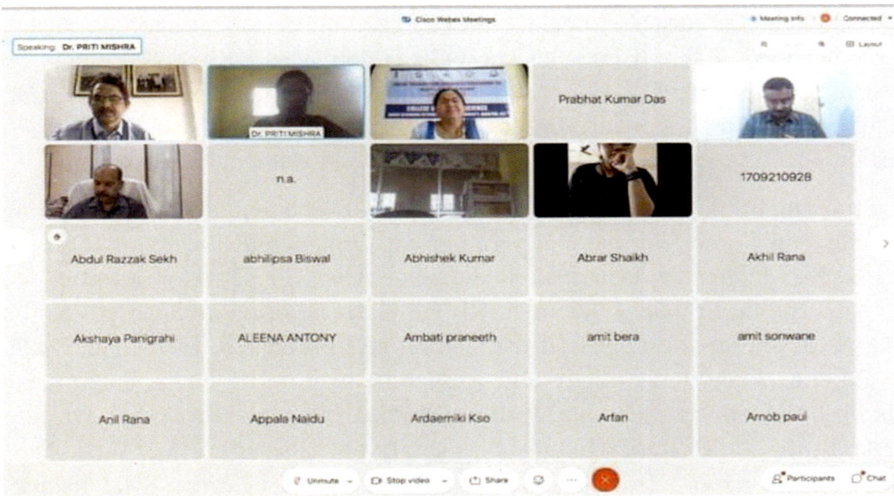

Fig. 18.5: Dr. B.P. Mohanty, ADG, Inland Fisheries, ICAR, New Delhi, delivered inaugural address on 19 June 2021. Other Delegates/Speakers present were Shri Bharat Singh, Director, Department of Fisheries, MP, Dr. Ambika Prasad Nayak, Scientist (Fishery Science) KVK, PURI, OUAT, Sakshigopal, Odisha, Dr. Madhuri Sharma, Coordinator of programme and farmers.

4. One-Day National Training and Awareness Programme on "Entrepreneurship Development Opportunities in Ornamental Fisheries- This programme was organised on 19 July, 2021 for students, entrepreneurs, farmers, fishing officials and others. The National Fisheries Development Board, Department of Fisheries, Govt. of India, under the Pradhan Mantri Matsya Sampada Yojana, funded this programme. Chief Guest of the programme Dr. B.P. Mohanty, ADG, Inland Fisheries, ICAR, New Delhi, delivered the special address in the inaugural session and emphasised that ornamental fisheries is an important sector with huge scope for expansion in both domestic and international markets. It could not only help in livelihood improvement for farmers and women self-help groups but also have immense opportunities for entrepreneurship and earning from high-value ornamental fisheries, also as envisaged in the PMMSY. The guest of honour **Shri Bharat Singh,** In-Charge Director, Department of Fisheries, MP, stated that the State Fisheries Department of the Govt. is doing a good job with the collaboration of ICAR institutes such as CIFA, CIFRI, etc. Special guest **Dr. S.K. Joshi,** Registrar, NDVSU, Jabalpur, stated that Madhya Pradesh with its highly productive climatic conditions provides scope for the development of ornamental fisheries. This sector assumes special significance due to its huge potential for providing employment to people hailing from rural areasand as foreign exchange earners. Special Guest and Keynote Speaker Dr. S.K. Swain, Director, ICAR-CIFA, Bhubaneswar, Odisha delivered a Keynote Lecture on the Entrepreneurship Development Opportunities in Ornamental Fisheries. Dr. Madhuri Sharma, Associate Professor Coordinator, stated that a total of 426 participants registered for this programme from 23 states in India.

The training programme covered most aspects of ornamental fishing including a lecture on Some Commercially Important Ornamental Fishes of India by Dr. B. K. Mahapatra, Principal Scientist, ICAR-CIFE, Kolkata Centre. Dr. Archana Sinha, Principal Scientist, CIFRI, Barrackpore, Kolkata delivered a talk on the status of ornamental fish marketing in India. Dr. B. Ahilan, Dean, Dr. M.G.R. Fisheries College and Research Institute, Ponneri, delivered a talk on Indigenous Ornamental Fishes of India: Status and Prospects. Dr. S.S. Dana, Professor, Fisheries Extension, Faculty of Fishery Science, WBUAFS, Kolkata, delivered a talk on entrepreneurship needs in ornamental fishing. Shri K.L. Maravi, In-Charge Joint Director Fisheries, Jabalpur Div. Jabalpur said that ornamental fish culture and exports areone of the recent areas of aquaculture diversification. He appreciated organising team for organising such a type of awareness programme for farmers, entrepreneurs, etc.

Fig. 18.6: Dr. B.P. Mohanty, ADG, Inland Fisheries, ICAR, New Delhi, delivered inaugural address on 19 June 2021. Other Delegates/Speakers present were Prof. R.P.S. Baghel, Dean, College of Fishery Science, NDVSU, Dr .S.K. Swain, Director, ICAR-CIFA; and Dr. Madhuri Sharma, Coordinator of the programme.

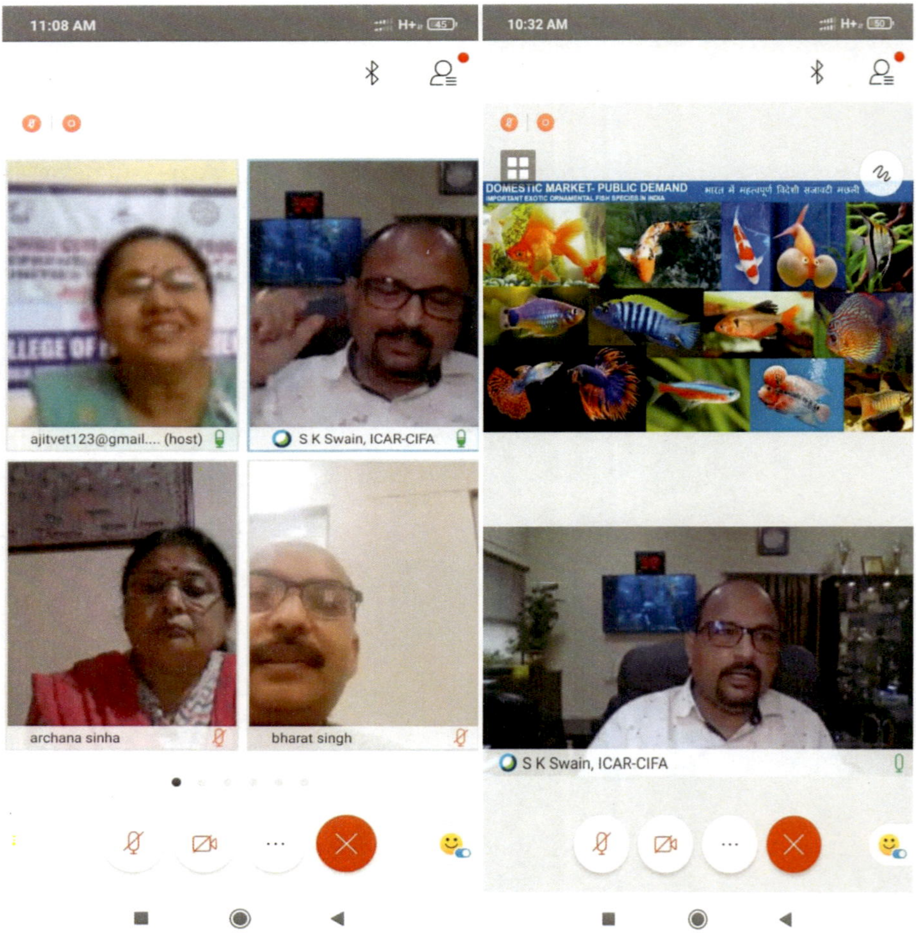

Fig. 18.7: Special Guest and Keynote Speaker, Dr. S.K. Swain, Director, ICAR-CIFA, delivered the keynote lecture on Entrepreneurship Development Opportunities in Ornamental Fisheries

5. Three day National Training on "Best Management Practices for Doubling the Income of Fish Farmers"- College of Fishery Science, NDVSU organized three day Physical Training on "Best Management Practices for Doubling the Income of Fish Farmers" from 4-6 August, 2021, for farmers and entrepreneurs. The National Fisheries Development Board, Department of Fisheries, Govt. of India, under the PMMSY, funded this programme. Prof. R.P.S. Baghel, Dean, College of Fishery Science and Dean Faculty, NDVSU, Jabalpur, delivered the welcome address. Dr. S. P. Tiwari, Hon'ble Vice-Chancellor, NDVSU, Jabalpur delivered the inaugural address. He said that with the proper management of feeding, disease, water and soil quality, our fish farmers can increase their income. Dr. Madhuri Sharma, Associate Professor Coordinator, stated that a total 25 participants were selected for this training programme from different districts of M.P. and Chattisgarh states.

The training covered most aspects of management practices for fish farming including a lecture on the importance of aquaculture, and modern techniques for aquaculture and transport management for fish seed and fish by Dr. Sunil Kumar Nayak, Scientist and Officer In-charge, ICAR-CIFE Powarkheda Centre, Dist. Hoshangabad, Madhya Pradesh. Dr. Madhuri Sharma delivered a talk on disease management practices for better fish production, pre- and -post stock management practices of composite fish culture and practical aspects of water and soil quality management for enhancement of fish production. Dr. Priti Mishra, Assistant Professor, delivered a talk on some commercially important freshwater species of fish and feeding management for enhancement of fish production. Dr. Sona Dubey, Assistant Professor delivered talk on water and soil quality management for enhancement of fish production and the practical aspects of transport management for fish seed and fish. Shri D.K. Jharia told about schemes of the Madhya Pradesh government for fish farmers. Shri Mukesh Kumar demonstrated practical aspects of disease management practices for better fish production. One field visit was organised for practical aspects of management practices for fish production.

Dr. M. K. Agnihotri, Controller of Examinations and Former ADG (HRD), ICAR, and Joint Commissioner, Govt. of India, New Delhi, addressed the participants as the Chief Guest of the Valedictory Programme. He said that proper management is very important for the better production of fish. Dr. S.P. Tiwari Hon'ble Vice-Chancellor, Nanaji Deshmukh Veterinary Science University, Jabalpur, addressed the participants as the Chairman of the Valedictory programme.

Fig. 18.8: The inaugural session of the awareness cum training programme on best management practices for doubling the Income of Fish Farmers, 4th to 6th August, 2021

Fig. 18.9: Practical demonstration of the management of hatchery units to farmers at Fish Farm Adhartal, Jabalpur

Fig. 18.10: Dr. M. K. Agnihotri, Controller of Examinations and Former ADG (HRD), ICAR, and Joint Commissioner, Govt. of India, New Delhi, addressed the participants as the Chief Guest and distributed a certificate in the valedictory session of the awareness cum training programme.

6. One Day Awareness cum Training Programme on "Importance of Fish in Nutrition": College of Fishery Science, NDVSU, organised a One Day Awareness cum Training Programme on "Importance of Fish in Nutrition" on September 20, 2021,at Bijadandi Block Mandla for entrepreneurs, farmers and others. The National Fisheries Development Board, Department of Fisheries, Govt. of India, under the PMMSY, funded this programme. This training was organised under the motivation and inspiration of the Hon'ble Vice-Chancellor, Nanaji Deshmukh Veterinary Science University, Dr. S.P. Tiwari, sir. Prof. R.P.S. Baghel, Dean, College of Fishery Science and Dean Faculty, NDVSU, Jabalpur, delivered a lecture on the importance of fish in nutrition. A total of 37 participants participated in this programme, which wasspecially organized on the occasion of Poshan Maah.

The training programme covered most aspects of the importance of fish in nutrition, including a lecture on the importance of fish in our life and preparation of fish pickles by Dr. Madhuri Sharma. Dr. Priti Mishra delivered a talk on the importance of fish in nutrition and preparation of fish cutlets.

Dr. S.K. Mahajan Associate and Dr. Sona Dubey delivered a talk on "Some commercially important edible fish. Dr. Madhuri Sharma, Dr. Priti Mishra and Shri Shivmohan demonstrated the practical aspect of fish Pickles and fish Cutlets.

Dr. S.P. Tiwari Hon'ble Vice-Chancellor, Nanaji Deshmukh Veterinary Science University, Jabalpur, addressed to the participants as the **Chairman of the Valedictory Programme and distributed fish seed and fish feed to fish farmers. Shri P.L. Yadav,** Chief Executive Officer, Janpad Panchayat, Bijadandi Block, Mandla, delivered a special speech on this occasion.

Fig. 18.11: The Valedictory session of the One Day Awarenessand Training Programme on "Importance of Fish in Nutrition

7. One-day National Awareness and Training Programme on "Disease and Health Management for Better Production of Fish"- College of Fishery Science, NDVSU organised this Awareness and Training Programme on February 9,2022. This programme was organised under the motivation and inspiration of Hon'ble Vice-Chancellor, NDVSU, Dr. S.P. Tiwari sir, with the able guidance of Dr. A. P. Dean, College of fisheries, and Dean Faculty, NDVSU, Jabalpur. Prof. Dr. M. K. Mehta, Dean Faculty, NDVSU, Jabalpur, delivered the welcome and inaugural address. Dr. B.P. Mohanty, ADG, Inland Fisheries, ICAR, New Delhi, Chief Guest, delivered the special address in the inaugural session. He said that fish is an important food, and with the proper management of feeding and disease, our fish farmers can increase their income and improve their socioeconomic status. Guest of Honour Shri Bharat Singh

Director, Department of Fisheries, Madhya Pradesh, stated that we want to make an Atmanirbhar. Madhya Pradesh. We are making efforts for this. A total of 502 participants registered for this programme, from 26 states of India.

The training focused on different aspects of fish health and disease management for sustainable aquaculture. A total of three lectures were delivered in one day. Dr. B. K. Das, Director, ICAR-CIFRI, Barrackpore, Kolkata delivered a talk on fish health management for ensuring better production. Dr. S.S. Mishra, Principal Scientist and Head, Fish Health Management Division, Bhubaneswar, Odisha delivered a talk on the bacterial diseases of freshwater fish and their management. Dr. Gadadhar Dash, Professor, Faculty of Fishery Sciences,WBUAFS, Kolkata, discussed parasitic diseases of freshwater fishand their remedial measures.

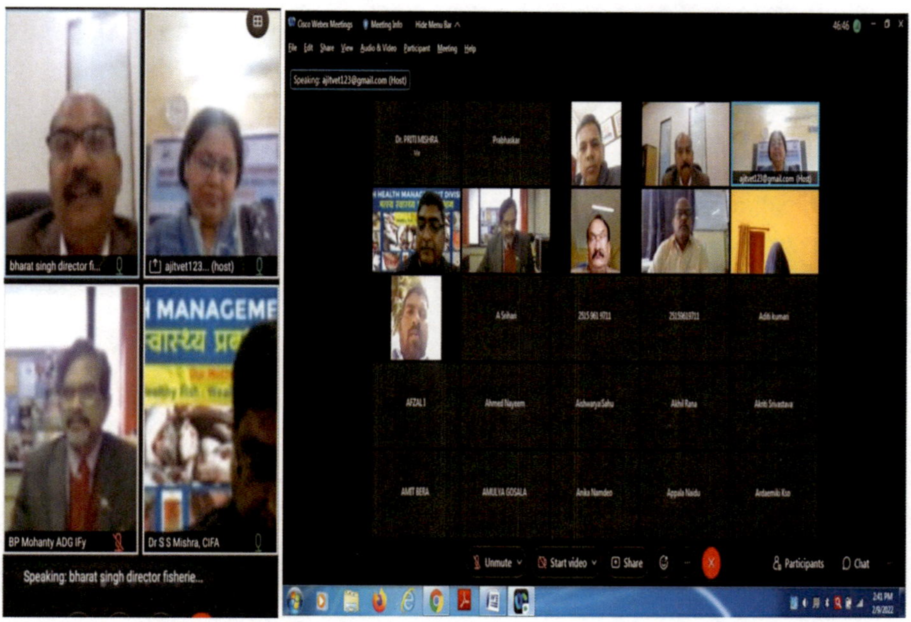

Fig. 18.12; Inaugural session of the Online National Awareness andTraining Programme on "Disease and Health Management for Better Production of Fish" Feruary 9., 2022. Dr. B.P. Mohanty, ADG, Inland Fisheries, ICAR, New Delhi, delivered the special address in the inaugural session and Guest of Honour Shri Bharat Singh, Director, Department of Fisheries. Govt. of M.P. was also present on this occasion

8. A one day training programme on "Water quality Management in Freshwater Aquaculture for Doubling the Income of Fish Farmers" was organised by College of Fishery Science (COFS), NDVSU, Jabalpur, on February 15. 2022. This programme was organised under the motivation and inspiration of Hon'ble Vice-Chancellor, Nanaji Deshmukh Veterinary Science University, Dr. S.P. Tiwari sir, and the able guidance of Dr. A. P. Dean, COFS, Jabalpur. Prof. Dr. S.P. Tiwari. The chief guest delivered the special address in the valedictory session. He said that this is a very important theme because water parameters such as pH, oxygen, etc. play an important role in fish farming. Coordinator Dr. Madhuri Sharma, Associate professor stated that this programme was organised for 35 fish farmers and entrepreneurs. This training programme received financial Support from NFDB under PMMSY scheme. Registrar of NDVSU Dr. S.K. Joshi, Director of Research Services, Dr. Madhu Swamy, and Director of Extension Education, Dr. S.K. Nayak, also graced the session.

The training focused on different aspects of water quality management for better production of fish. A total of 3 lectures were delivered in one day. Dr. Madhuri Sharma,Associate Professor, delivered a talk on the management of chemical parameters for enhancement of fish production. Miss Priyanka Gautam, a teaching associate, delivered a talk on the management of physical and biological parameters of water for better production. Shri Shiv Mohan Singh, Teaching Associate, demonstrated practical aspects of the management of chemical parameters for enhancement of fish production. Shri Mukesh Kumar, a teaching associate, demonstrated the practical aspects of management of physical and biological parameters of water for better production.

Fig. 18.13: Inaugural session of the One Day Physical Training Programme on Water Quality Management in Freshwater Aquaculture for Doubling the Income of Fish Farmers, February 15,2022. Hon'ble Vice- Chancellor, NDVSU Prof. Dr. S.P. Tiwari, Chief Guest, delivered the addresson this occasion.

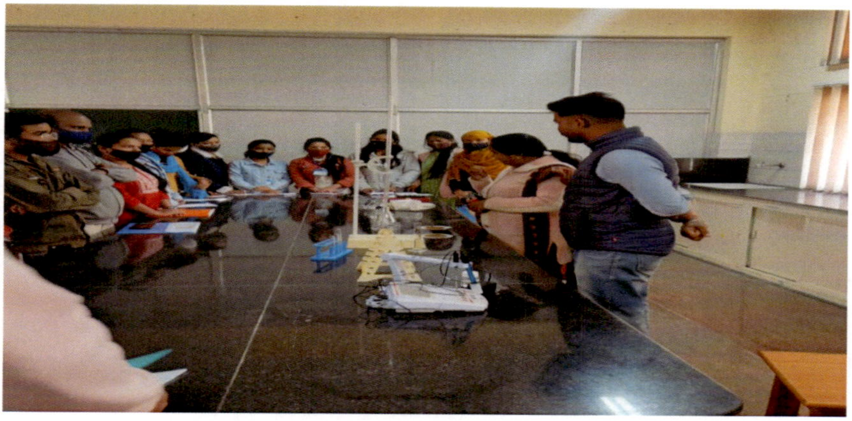

Fig. 18.14: Practical demonstration of Water quality analysis

Fig. 18.15: Valedictory session of the Physical Training Programme on Water Quality Management in Freshwater Aquaculture for Doubling the Income of Fish Farmers. Hon'ble Vice-Chancellor, NDVSU Prof. Dr. S.P. Tiwari, Chief Guest, delivered the special address on this occasion.

9. One-day National Awareness and Training Programme on Feed Formulation and Management in Freshwater Aquaculture- College of Fishery Science, NDVSU celebrated National Science Day and organised a one day National Awareness and Training Programme on Feed Formulation and Management in Freshwater Aquaculture on February 28,, 2022. This programme was organised under the motivation and inspiration of Hon'ble Vice-Chancellor, Nanaji Deshmukh Veterinary Science University, Dr. S.P. Tiwari sir, and with the able guidance of Dr. A. P. Dean, College of Fishery and Dean Faculty, NDVSU, Jabalpur. Prof. Dr. M. K. Mehta, Dean Faculty, NDVSU, Jabalpur, delivered the welcome and inaugural address. Dr. B.P.

Mohanty, ADG, Inland Fisheries, ICAR, New Delhi, Chief Guest, delivered a special address in the inaugural session. February 28 was celebrated National Science Day and the theme of this progamme is very relevant. He said that fish feed is essential to the better growth of fish, and proper management of feeding our fish farmers can increase their income. There is a need to develop good quality and cheapest feed for fish culture, because at present, feed costs are very high.The guest of honour Shri Bharat Singh, Director, Department of Fisheries, Madhya Pradesh, stated that we are making efforts in this direction. We are establishing some feed mills in some districts of M.P. Organising Secretary Dr. Madhuri Sharma, Associate Professor, stated that atotal of 423 participants registered for this programme from 26 states in India. This programme has received financial assistance and support from NFDB Hyderabad under the PMMSY Scheme.

The training focused on different aspects of fish health and disease management for sustainable aquaculture. A total of 3 lectures were delivered in one day. Dr. K. K. Jain, Retired Professor/Principal Scientist, Fish and Fisheries Science, ICAR-CIFE, Mumbai, delivered a talk on the importance of fish feed and its type. Dr. K. N. Mohanta, Principal Scientist, ICAR-CIFA, Bhubaneswar, Odisha, delivered a talk on the nutritional requirements and feeding of freshwater fish. Dr. Sikendra Kumar, Scientist, ICAR-CIFE, Mumbai, discussed feed preparation.

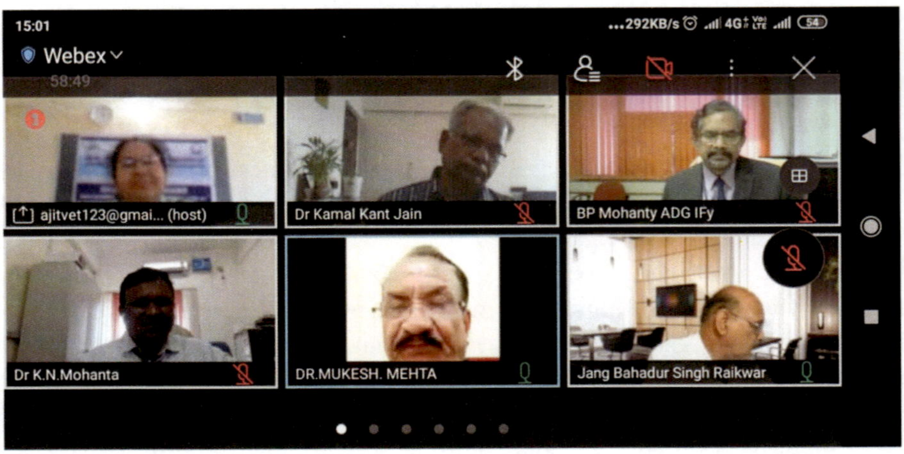

Fig. 18.16: The inaugural session of the one-day training programme on feed formulation and management in freshwater aquaculture onFebruary 28,, 2022. This programme funded by the NFDB under the PMMSY. Dr. B.P. Mohanty, ADG, Inland Fisheries, ICAR, New Delhi, Chief Guest, delivered the special address in the inaugural session and Guest of HonourShri Bharat Singh, Director, Department of Fisheries, Govt. of M.P.,was also present on this occasion.

10. One-day National Awareness and Training Programme on Pradhan Mantri Matsya Sampada Yojana (PMMSY)- College of Fishery Science, NDVSU organised a one-day National Awareness and Cumulative Training Programme on PMMSY on March 14, 2022. This programme was organised under the motivation and inspiration of Hon'ble Vice-Chancellor, Nanaji Deshmukh, Veterinary Science University, Dr. S.P. Tiwari sir, and the able guidance of Dr. A. P. Dean, College of fisheries and Dean Faculty, NDVSU, Jabalpur, Prof. Dr. M. K. Mehta, Dean Faculty, NDVSU, Jabalpur, delivered the welcome and inaugural address. Dr. Joykrushna Jena, DDG, Fisheries, ICAR, New Delhi, Chief Guest, delivered the special address in the inaugural session. He said that PMMSY is the most important and biggest scheme in the fishing sector. The PMMSY scheme has several opportunities for fish farmers and entrepreneurs. Guest of Honour Dr. L. M. Murthy, Senior Executive Director, NFDB, Hyderabad, said that there is a need for proper skill development for better production of fish. We can develop the skills of fish farmers and entrepreneurs through wareness and training programmes, etc., and increase their income, and improve their socioeconomic status. Special Guest and Speaker, Dr. S.K. Swain, Director, ICAR-CIFA, Bhubaneswar, Odisha, delivered a special address in the inaugural session. A total of 377 participants registered for this programme from 23 states in India. This awareness programme has received financial assistance and Support from NFDB Hyderabad.

The training focused on different aspects of PMMSY. A total of three lectures were delivered in one day by Dr. S.K. Swain, Director, ICAR-CIFA addressing the importance of aquaculture and fisheries for food and livelihood, Dr. S.N. Ojha, Principal Scientist and Head, FEES Division, ICAR-CIFE, Mumbai, delivered a talk on PMMSY, and Mr. Priyak Shrivastav, Department of Fisheries, Jabalpur, delivered a talk on schemes, loans and other beneficiary programmes for fish farmers. The training was well coordinated by Dr. Madhuri Sharma, Associate Professor. Dr. Priti Mishra acted as Co-oordinator. Mrs. Shashi Prabha Dhurve, In-Charge Joint Director Department of Fisheries, Jabalpur Division, was also present in the inaugural session. Dr. Hari R., Assistant Professor, Department of Veterinary and Animal Husbandry Extension Education, College of Veterinary Science and A.H., Jabalpur, proposed a vote of thanks.

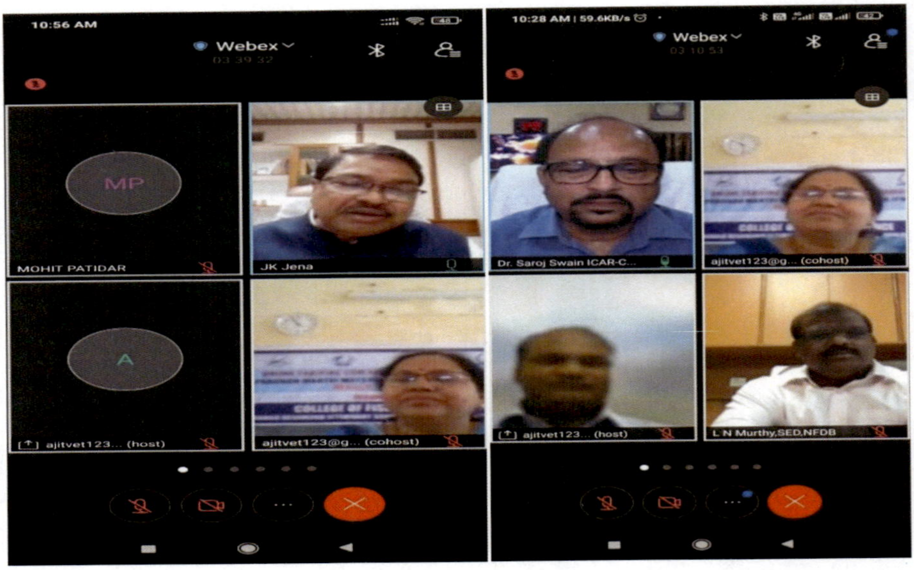

Fig. 18.17: The inaugural session of the Online National Awareness andTraining Programme on PMMSY on March 14, 2022. Dr. Joykrushna Jena, DDG, Fisheries, ICAR, New Delhi, Chief Guest, delivered the special addresson this occasion.The webinar scene also shows S.K. Swain, Director, ICAR-CIFA; Dr. L. M. Murthy, Senior Executive Director, NFDB, Hyderabad; Dr. A. P. Dean, College of Fisheries NDVSU, Jabalpur, and Dr. Madhuri Sharma, Coordinator of programme.

11. Awareness and Training Programme on Fish Farmer Producer Organization (FFPO) under PMMSY was organised on August 3,2022 organised by the Regional Directorate, NCDC, Bhopal, the College of Fishery Science, Jabalpur and the Department of Fisheries and the Ministry of Animal Science and Dairy at the College of Veterinary Science and AH, NDVSU, Jabalpur. More than 70 farmers participated in this program. Different experts delivered a lecture on FFPO Concept.

Fig. 18.18: The inaugural session of Awareness and Training Programme: Hon'ble Vice-Chancellor, NDVSU Prof. Dr. S.P. Tiwari, Chief Guest delivered the address and Regional Director, NCDC, Bhopal Shri Shailendra Singh also addressed the farmers onthis occasion.

Fig. 18.19: The Valedictory session of the awareness and training programme on FPPO on August 3, 2022

Finally, I am thankful to all the participants who registered in awareness and training programmes on different aspects of PMMSY. I am especially grateful to Prof. (Dr.) S.P. Tiwari, Hon'ble Vice-Chancellor, Nanaji Deshmukh Veterinary Science University, Jabalpur, Dr. R.P.S. Baghel, and Dr. A.P. Sing Dean, College of Fishery Science, for their motivation, guidance, and continuous support in organising these awareness and training programmes. I am also thankful to Prof. Dr. M.K. Mehta, Dean of Faculty and Dr. S.K. Joshi, Registrar of NDVSU, for their support and guidance. I am also thankful to NFDB Hyderabad for providing financial support to 12 awareness and training programmes under PMMSY. Special votes of thanks are presented to all the speakers for providing their valuable time. I am especially grateful to the Regional Director, NCDC, Bhopal Shri Shailendra Singh, for organising a collaborative programme with us. I am especially grateful to Dr. Joykrushna Jena (DDG, Fisheries, ICAR, New Delhi) and Dr. B.P. Mohanty (ADG, Inland Fisheries, ICAR, New Delhi). I am also thankful to Course Co-Coordinator Dr. Priti Mishra and Technical Committee Members of different programmes.

19

Capacity Building Initiatives of the Indian Council of Agricultural Research in Fisheries and Aquaculture During Azadi ka Amrit Mahotsav

Yasmeen Basade, Shubhadeep Ghosh and Bimal Prasanna Mohanty

Fisheries Science Division, Indian Council of Agricultural Research, KAB II Pusa, New Delhi

---

India's journey through the 75 years of independence and its splendid achievements are acclaimed by the Government of India through its initiative, Azadi Ka Amrit Mahotsav. Since independence, the nation has traversed a great distance and achieved feats in various spheres, such as agriculture, science, technology, art and culture, to name a few. Azadi Ka Amrit Mahotsav commemorates 75 glorious years of progressive India and its rich history, diverse population, magnificent culture and great achievements. The Hon'ble Prime Minister of India, Shri Narendra Modi's, through this campaign, envisioned the foundation of a self-reliant India, and he said, "*The Azadi Ka Amrit Mahotsav means elixir of energy of independence; elixir of inspirations of the warriors of freedom struggle; elixir of new ideas and pledges; and elixir of Aatmanir bharta. Therefore, this Mahotsav is a festival of awakening the nation a festival of fulfilling the dream of good governance, and a festival of global peace and development*"[1].

The government of India, to celebrate the 75th anniversary of India's independence, organised a series of events under the campaign Azadi Ka Amrit Mahotsav. The Mahotsav is being celebrated as a Jan-Utsav in the spirit of Jan-Bhagidari. The official journey of Azadi ka Amrit Mahotsav commenced on March 12, 2021 which started as a 75-week countdown to our 75th anniversary of independence and will end post a year on August 15, 2023[1]. The events were organised under five themes viz., Freedom Struggle, Ideas@75, Actions@75, Achievements@75, and Resolve@75 as given below:

Table 19.1: Themes of Azadi Ka Amrit Mahotsav[2]

Freedom Struggle	Resolve@75	Ideas@75	Actions@75	Achievements@75
Commemorating milestones in history, unsung heroes, etc.	Reinforcing commitments to specific goals and targets	Celebrating the ideas and ideals that have shaped India	Highlighting steps being taken to implement policies and actualize commitments.	Showcasing evolution and progress across different sectors

(Source: https://amritmahotsav.nic.in/themes.htm website visited on 11-03-2024)[2]

To commemorate Azadi ka Amrit Mahotsav, the Indian Council of Agricultural Research (ICAR) identified activities and events for each of the 75 weeks, commencing from March 12, 2021, the first week and up to August 15, 2022, the 75th week, earmarking the 75th India's Independence Day.. Activities were planned involving community participation to generate awareness about its scientific achievements and technological developments. Major activities coordinated included: 75 mass awareness campaigns on specific themes and commodities; 75 lectures by eminent scholars; documentation of 75,000 success stories of doubling farmers income; and publication of Indian Agriculture: A Saga of Success: Role of ICAR in Agricultural Revolutions[3].

Azadi Ka Amrit Mahotsav has later been extended up to August 15, 2023, to further boost this people's movement by focusing on critical areas of cultural and social development and identifying new themes, Theme 2.0, aligned with the 'Panch Pran' announced by Hon'ble Prime Minister: Women and Children, Tribal Empowerment, Water, Cultural Pride, Lifestyle for Environment (LiFE), Health and Wellness, Inclusive Development, Aatmanirbhar Bharat, and Unity[4]. These are as given below.

Table 19.2: Themes 2.0 of Azadi Ka Amrit Mahotsav

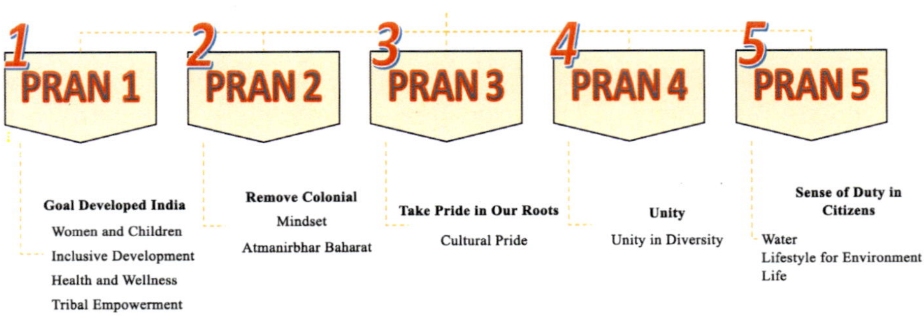

(Source: https://amritmahotsav.nic.in/themes-2-0.htm website visited on 11-03-2024)[4]

The mass awareness campaigns were conceived as national campaigns and sector-specific campaigns. The events were envisaged with emphasis on the accomplishment of the objectives of Amrit Kaal with high Jan Bhagidari and public participation. National campaigns were organised, coinciding with special occasions of national and international importance, engaging all the ICAR institutes, Krishi Vigyan Kendra's (KVKs), State Agriculture Universities (SAUs), farmers, and public leadership. Sector-specific campaigns were focused on the crop sector, horticulture sector, natural resource management (NRM), agricultural engineering sector, animal sector, and the fisheries sector. All these sectoral campaigns were organised by the respective specialised ICAR institutions, ICAR-All India Coordinated Research Project, the cooperating centres of SAUs and KVKs. Seminars/webinars, kisan gosthies, field days, meetings and outreach programmes were conducted to create awareness amongst various stakeholders to contribute towards the process of nation building. The details of national and sector specific campaigns are presented below[3].

National Campaigns: The National Campaigns were structured to be organised coinciding with special occasions of national and international importance, involving all the ICAR Institutes, KVKs, SAUs, farmers, public leadership, and others. The details of 22 national campaigns are as follows:

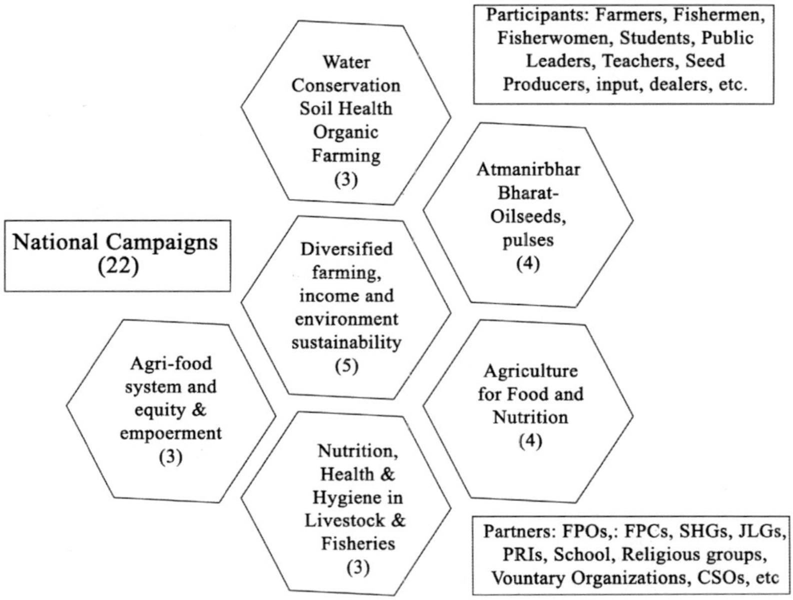

Fig. 19.1: National Campaigns (Source: https://icar.org.in/sites/default/files/Circulars/DARE-ICAR%2075%20Years%20%282%29.pdf website visited on 11-03-2024)

Sector-Specific Campaigns: In addition to national campaigns, the sector-specific campaigns were organised by involving the specialised ICAR Institutions, ICAR-All India Coordinated Research Project, Cooperating Centres of SAUs and KVKs. Webinars, kisan gosthies, field days, meetings, and outreach programmes were conducted to create awareness amongst various stakeholders. These campaigns were organised for the Crop, Horticulture, Natural Resource Management, Agricultural Engineering, Animal and Fisheries Sectors. The details of Fisheries sector campaigns are given below.

Fisheries Sector: Nine awareness campaigns under different thematic areas were organized.

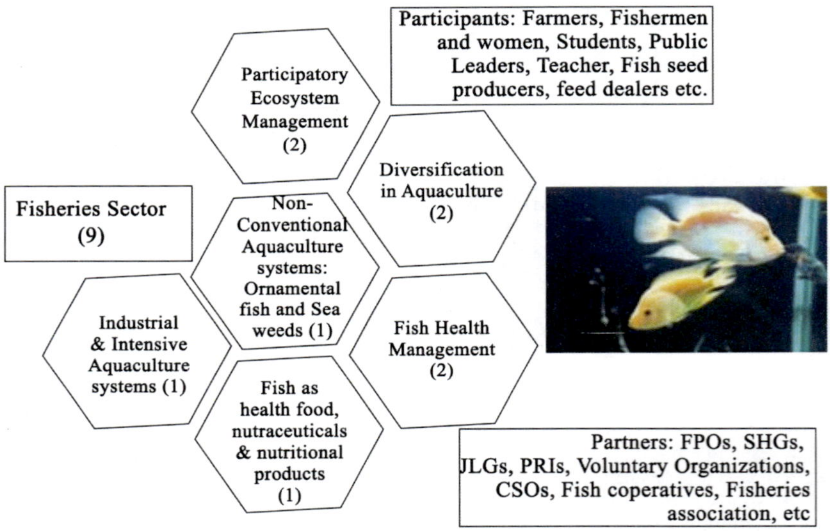

Fig. 19.2: Fisheries Sector Campaigns (Source: https://icar.org.in/sites/default/files/Circulars/DARE-ICAR%2075%20Years%20%282%29.pdf website visited on 11-03-2024)

The national campaigns were organized by all the ICAR research institutes across the disciplines. Fisheries Sector Campaigns were chiefly organised by all the ICAR Fisheries Research Institutes and also by the Institutes and KVKs undertaking integrated farming with fish as per the themes and schedule ascertained by the ICAR.

The campaigns related to the Fisheries Sector were coordinated by the Fisheries Science Division of ICAR and implemented through its five resource specific fisheries research institutes viz., ICAR-Central Marine Fisheries Research Institute (CMFRI), ICAR-Central Inland Fisheries Research Institute (CIFRI), ICAR-Central Institute of Fisheries Technology (CIFT), ICAR-

Central Institute of Freshwater Aquaculture (CIFA), ICAR-Central Institute of Brackishwater Aquaculture (CIBA); a Deemed University, ICAR-Central Institute of Fisheries Education (CIFE); a bureau, ICAR-National Bureau of Fish Genetic Resources (NBFGR); and a directorate, ICAR-Directorate of Coldwater Fisheries Research (DCFR); which are in turn supported by a wide network of regional, research and field centres. In addition to the ICAR-Fisheries Research Institutes, the other ICAR Institutes were: ICAR-Central Coastal Agricultural Research Institute, Goa; ICAR-Mahatma Gandhi Integrated Farming Research Institute (MGIFRI), Motihari; ICAR-Research Complex for Eastern Region (RCER), Patna; ICAR-Research Complex for North East Hill Region (RCNEH), Umiam; ICAR-Indian Institute of Water Management (IIWM), Bhubaneswar; ICAR-National Bureau of Agriculturally Important Microorganisms (NBAIM), Mau Nath Bhanjan and ICAR-National Research Centre for Integrated Pest Management (NRCIPM), New Delhi. These campaigns were implemented by the institutes directly or in collaboration with the related state departments and organisations, involving the local masses, students, individual farmers, farming communities, and other stakeholders. The national and fisheries sector-specific Campaigns organised are detailed below.

NATIONAL CAMPAIGNS

Food and Nutrition for Farmers
(August 26, 2021)

Gosthi on 'Food and Nutrition for Farmers' organized at Lorna Reservoir, Simdega district, Jharkhand by ICAR-CIFRI (Source: ICAR-CIFRI)

PROGRAMMES ORGANIZED & PARTICIPANTS

Gosthi: 5; Seminar/Webinar: 4; Training: 2
Participants: 523

Waste to Wealth
(October 12, 2021)

Awareness programme on 'Waste to Wealth' and swachhta campaign organized at Government Inter College, Talla Ramgarh, Nainital by ICAR-DCFR (Source: ICAR-DCFR)

PROGRAMMES ORGANIZED & PARTICIPANTS

Mass awareness: 5
Participants: 615

Poshan Vatika and Tree Plantation
(September 17 2021)

Poshan Vatika and Tree plantation program at ICAR-CIBA, Kakdwip Research Centre, West Bengal (Source: ICAR-CIBA)

Millet's products being distributed to women farmers and girls at ICAR-CIFRI, Barrackpore (Source: ICAR-CIFRI)

Farmers being explained the concept of 'Nutri-Garden and millets' at ICAR-CIFE, Rohtak Centre (Source: ICAR-CIFE)

PROGRAMMES ORGANIZED & PARTICIPANTS

Plant saplings/ Millet products distributed: 4237
Participants: 2535

Mahila Kisan Diwas
(October 15, 2021)

Mahila Kisan Divas celebration organized at Agatti, Lakshdweep by ICAR-NBFGR

PROGRAMMES ORGANIZED & PARTICIPANTS
Lecture: 3; Mass awareness: 2; Gosthi: 4; Seminar/Webinar: 5; Training: 2
Participants: 731

World Food Day
(October 16, 2021)

ICAR-CIFA and KVK, Khordha organized World Food Day and felicitated fish farmers for their contribution (Source: ICAR-CIFA)

PROGRAMMES ORGANIZED & PARTICIPANTS
Mass awareness:1; Seminar/Webinar: 2
Participants: 377

World Soil Day
(December 05, 2021)

Sensitization programme organized for farmers at Renukoot in Uttar Pradesh by ICAR-NBFGR (Source: ICAR-NBFGR)

PROGRAMMES ORGANIZED & PARTICIPANTS
Invited talk: 5; Mass awareness: 4
Participants: 358

Annadata Devo Bhava
(April 23-24, 2022)

On the occasion webinar on 'Natural fish farming' organized by ICAR-CIFRI, Barrackpore (Source: ICAR-CIFRI)

PROGRAMMES ORGANIZED & PARTICIPANTS
Mass awareness: 2; Seminar/Webinar: 1
Participants: 537

Kisan Bhagidari, Prathmikta Hamari

(April 25-28, 2022)

Awareness programme organized for Women Self Help Group Members by ICAR-CIFA (Source: ICAR-CIFA)

PROGRAMMES ORGANIZED & PARTICIPANTS

Invited talk: 1; Mass awareness: 4; Seminar/Webinar: 2

Participants: 789

Har Ghar Tiranga

(August 13-15, 2022)

Scientists of ICAR-CIFRI visited Khalsi Wetland, Haringhata, Nadia and distributed flags to Khalsi Udbastu Samabay Samiti fishermen society members to promoted the concept of Har Ghar Tiranga. (Source: ICAR-CIFRI)

PROGRAMMES ORGANIZED & PARTICIPANTS

Mass awareness: > 23; Meet: 2

Participants: > 846

Fig. 19.3: Snapshots of National Campaigns organized by the ICAR Institutes

FISHERIES SECTOR CAMPAIGNS

Ecosystem Management for Sustainable Fisheries

(July 03-10, 2021)

Awareness programme organized for fisherwomen of villages Pamban and Thoppukadu, district Ramanathapuram by ICAR-CMFRI, Regional Station, Mandapam, Tamil Nadu (Source: ICAR-CMFRI)

Kisan Gosthi organized on 'Reservoir Fisheries Management for Sustainable Development' at Manchanabele reservoir, Ramanagara district, Karnataka by ICAR-CIFRI (Source: ICAR-CIFRI)

PROGRAMMES ORGANIZED & PARTICIPANTS

Gosthi: 13; Seminar/Webinar: 29; Training: 2

Participants: 4074

System Diversification in Aquaculture

(August 08-September 01, 2021)

Awareness cum training programs organized by ICAR-NBFGR with its project partners Coastal & Marine Biodiversity Centre, Airoli, Maharashtra (Source: ICAR-NBFGR)

PROGRAMMES ORGANIZED & PARTICIPANTS

Gosthi: 6; Seminar/Webinar: 15;
Training: 7
Participants: 1775

Diversification in Aquaculture

(March 10, 2022)

Webinar on 'Diversification in Aquaculture' organized by ICAR-CMFRI (Source: ICAR-CMFRI)

PROGRAMMES ORGANIZED & PARTICIPANTS

Gosthi: 2; Seminar/Webinar: 16
Participants: 2126

Antimicrobial Resistance in Fish

(November 18- 24, 2021)

International Webinar on 'Antimicrobial Resistance in Fish' organized by ICAR-CIFT (Dr. C.N. Ravisankar, Directors, ICAR-CIFT, Kochi; Dr. J.K. Jena, DDG (FS), ICAR, New Delhi; Prof. Stephen Rimmer, University of Bradford and Prof. Till Bachmann, University of Edinburgh seen in photograph, L-R) (Source: ICAR-CIFT)

Seminar on "Recent advances on AMR surveillance" organized by ICAR-NBFGR(Source: ICAR-NBFGR)

Farmer-Scientist Meet organized by ICAR-DCFR at villages Harinagar, Uttarakhand to make farmers aware about the banned antibiotics and chemicals in aquaculture (Source: ICAR-DCFR)

PROGRAMMES ORGANIZED & PARTICIPANTS

Gosthi: 16; Seminar/ Webinar: 18;
Training: 1
Participants: 4898

Emerging Aquaculture Systems and Practices

(July 11, 2022)

Farmer-scientist interaction meeting organized by ICAR-CIFRI in collaboration with Moyna Ramkrishnayan Association at Moyna fisheries hub, East Medinipur, West Bengal (Source: ICAR-CIFRI)

Workshop on 'Small is big: Enhancing productivity in small scale aquaculture' organized by ICAR-CIFE, Kolkata Centre (Source: ICAR-CIFE)

Conference on "Emerging Aquaculture Systems and Practices" and Farmers' Meet organized by ICAR-DCFR (Source: ICAR-DCFR)

PROGRAMMES ORGANIZED & PARTICIPANTS

Gosthi: 5; Seminar/Webinar: 25; Training: 26; Exhibition: 1

Participants: 3242

Non-conventional Aquaculture Systems

(July 27, 2022)

Farmers' interaction cum demonstration programme on 'Cage Aquaculture of brackishwater fishes in the coastal waters' organized by ICAR-CIBA at Thonirevu village, Pulicat, Thiruvallur district of Tamil Nadu (Source: ICAR-CIBA)

Awareness program organized by ICAR-NBFGR at 'Adarsh Hatchery' situated in village Kataundhi, Sonbhadra district, Uttar Pradesh (Source: ICAR-NBFGR)

Kisan gosthi on 'Non-conventional aquaculture systems' organized by ICAR-CMFRI, Regional Stations, Veraval (Source: ICAR-CMFRI)

PROGRAMMES ORGANIZED & PARTICIPANTS

Gosthi: 11; Seminar/Webinar: 14; Training: 2

Participants: 2604

Fish Health Management in Aquaculture
(July 29, 2022)

Online talk organized by ICAR-CIBA on "Emergent diseases and efforts toward developing antiviral therapy in shrimp" by Dr Arun K. Dhar, Professor & Director, Aquaculture Pathology Laboratory, School of Animal & Comparative Biomedical Sciences, the University of Arizona, Tucson, USA (Source: ICAR-CIBA)

PROGRAMMES ORGANIZED & PARTICIPANTS
Gosthi: 1; Seminar/Webinar: 12
Participants: 1691

Participatory Ecosystem Management for Sustainable Fisheries and Biodiversity Conservation
(August 04, 2022)

Webinar on 'Participatory Management for Sustainable Fisheries and Biodiversity Conservation' organized by ICAR-NBFGR (Source: ICAR-NBFGR)

PROGRAMMES ORGANIZED & PARTICIPANTS
Gosthi: 3; Seminar/Webinar: 22
Participants: 3315

Fish for Health and Prosperity/ Fish as Health Food
(July 15 - August 10, 2022)

Awareness was created regarding importance and role of fish in human nutrition by ICAR-CIFT at Govt. Boys High School, Gajapathinagaram, Vizianagaram; and Cheredev Macchimar Seva Santha, Ghansoli (L-R) (Source: ICAR-CIFT)

PROGRAMMES ORGANIZED & PARTICIPANTS
Gosthi: 56; Seminar/Webinar: 32; Training: 9
Participants: 40865

Fig. 19.4: Snapshots of Fisheries Sector Campaigns organised by the ICAR Institutes.

Conclusion

The 75th year of India's independence was celebrated as Azadi ka Amrit Mahotsav. Various activities were planned by ICAR and organised by the Fisheries Science Division, ICAR. These included awareness campaigns and lectures by eminent persons on the thematic areas discussed above for reaching out to maximum stakeholders, particularly farmers, on various new developments in different sectors of fisheries. Rural youths, farmers, school and college students, entrepreneurs, and people from all walks of life attended the various campaigns organised across the country. In, all more than sixty-two thousand participants attended the National Campaigns and the Fisheries Sector Campaigns from the different states across the country[5].

Acknowledgements

The authors thankfully acknowledge the Secretary, DARE and DG, ICAR, New Delhi; DDG (Fisheries Science), ICAR, New Delhi and the Indian Council of Agricultural Research (ICAR), New Delhi, for planning the campaigns and activities of Azadi Ka Amrit Mahotsav; and also, to the Director's and staff members of all the partaking institutes for their active participation.

References

Azadi ka Amrit Mahotsav, About. Available at: https://amritmahotsav.nic.in/about.htm (accessed on March 11, 2024).

Azadi ka Amrit Mahotsav, themes. Available at: https://amritmahotsav.nic.in/themes.htm (accessed on March 11, 2024).

Mishra, J.P. 2021. Azadi Ka Amrut Mahotsav, Commemoration of the 75th Year of India's Independence, Programmes and Activities. Publ. Indian Council of Agricultural Research,Department of Agricultural Research and Education, Krishi Bhawan, New Delhi. Available at https://icar.org.in/sites/default/files/Circulars/DARE-ICAR%20 75%20Years%20%282%29.pdf (accessed on March 11, 2024).

Azadi ka Amrit Mahotsav, Themes 2.0. Available at: https://amritmahotsav.nic.in/themes-2-0. (htm (accessed on 11, March 2024).

Promoting Fisheries for Food, Nutrition, and Livelihood Security. A compilation of activities of Azadi Ka Amrit Mahotsav, organised by the Fisheries Science Division, Indian Council of Agricultural Research (ICAR), New Delhi, pp. 70 (to be published).

20

Capacity Building on Sustainable Fisheries for Professionals from Mekong Ganga Cooperation (MGC) Countries A Way Forward Towards Strengthening of Neighbourhood and Achieving Sustainable Development Goal (SDG)

A.K.Sahoo, B.K.Das, D.K.Meena, S.M. Nair, P. DebRoy, Mishal P. and B.P. Mohanty

ICAR-Central Inland Fisheries Research Institute, Barrackpore, Kolkata
Fisheries Science Division, Indan Councal of Agricultural Reasearch
KAB II, Pusa, New Delhi

Introduction

Fisheries and aquaculture sector have been increasingly recognized for their essential contribution to global food security and nutrition. In 2020, total global catches in inland waters were 11.47 million tonnes and India stands top in the world with 1.8 million tonnes. India has witnessed remarkable growth in fish production in the last couple of years and stands second in world aquaculture production. The total fish production during 20-21 is estimated at 14.7 million metric tonnes (MMT) with a contribution of 11.25 MMT from inland sector and 3.48 MMT from marine sector. During FY 2020-21, export of marine products stood at 1.15 MMT and valued at Rs. 43,717.26 crores despite the market uncertainties raised by the Covid-19 pandemic outbreak. The inland open water resources of India are the life line of country"s vast population and provide nutritional and livelihood security besides other innumerable economic, social and ecological services. In addition to India, countries of Mekong cooperations including Myanmar, LaoPDR, Vietnam,

Cambodia and Thailand contribute significantly to the Asian Fish production. As per FAO (2022) report, a number of Asian countries majorly contribute to the Inland capture fisheries production including India, Bangladesh, Myanmar and Cambodia have reported a steady or decline in fish catch, since the last few decades. Though there are several potential threats behind this declining fish production trend, major is linked to water availability, pollution, overfishing and climate change. This necessitates proper resource management, based on sound and informed decision making and technological interventions for protecting the goods and ecological services to sustain the fisheries.

Sustainability in fisheries management requires enhancement of scientific and technological capacities (Espinoza-Tenorio *et al.*, 2011). Capacity building in natural resource management is the way in which the capability of a country in achieving the efficient and effective pursuit of environmental goals is achieved in co-operation with other nations (Chircop, 1998). In environmental development work, capacity building has a job-specific focus that aims to fill knowledge gaps and provide skills (UNCED, 1992). In spite of the necessity of capacity building in sustainable fisheries resource management, there are economic and political challenges in it (FAO, 1995). Overcoming these challenges becomes important in developing countries because much of the global fish production is harnessed from such nations. Globally, commercial fisheries started booming from 1970s; however, the fisheries set-ups in many places have had limited training and institutional capacities ((Espinoza-Tenorio *et al.*, 2011). In view of the importance of international capacity building for fostering collaborative transboundary fisheries management in the inland sector, the International Workshop-cum-Training programme with four Mekong Co-operation countries namely Cambodia, Myanmar, Lao PDR and Vietnam was organized at ICAR – Central Inland Fisheries Research Institute in Barrackpore, Kolkata in West Bengal, India to train the officials and researchers of the four countries through imparting knowledge and skill in addition to field-cum-exposure visits in various fisheries centres in West Bengal (Fig 1).

Fig. 20.1: Group photo of Mekong-Ganga Cooperation Country trainees on Sustainable fisheries

There has to be a concerted attempt to build capacity among fishery/ aquaculture professionals for Inland open water fisheries management. The capacity programme aims a) to review the current management practices and critically identify and address the constraints, risks and other problems and propose possible improvements b) to demonstrate relevant technologies/skills that are essential for fisheries enhancement/CBF/enclosure culture practices and to generate trained manpower c) to disseminate principles and practices of inland open-water fisheries management in the context of food and nutritional security. These aims were achieved through series of expert lecturers, field visits, fishermen interactions and hands on training on site and in laboratory. The training programme was participated by 24 professional representing from Vietnam, Cambodia, LaoPDR, and Myanmar for a period of 14 days. Major significant areas those need for further strengthening among the MGC counties were highlighted here and are given.

Selective breeding and exchange of fish genetic materials

Selective breeding in aquaculture is the best tool to improve genetic status of the selected fish species in direction of enhanced growth, disease resistance, quality of flesh and therefore the increase in production and productivity. In India, through selective breeding program improved variety of Rohu *Labeo rohita* has been developed by ICAR-Central Institute of Freshwater Aquaculture (ICAR-CIFA) and popularly known as "Jayanti" (Fig 2). For developing the genetic improved Jayanti Rohu, the base population was established from five riverine sources. The average realized selection response of harvest body weight after eight generations of selection was 18.3% per generation.

Multilocational field testing as well as on-farm trials in different parts of India showed significant growth improvement. The improved Jayanti species has been disseminated to 16 states farms and different hatcheries of India (Rasal *et al*. 2017). The result on the growth of Jayanti rohu at different locations was significantly higher than the normal. In addition to growth, disease resistance against *Aeromonas hydrophila* has been the second trait of selection to rohu breeding program. The improved rohu also proved to be feed neutral and effective under all culture practices which have enhanced the acceptance level. Impact assessment study further confirmed superiority of Jayanti strain of rohu over the normal rohu.

Fig. 20.2: Genetically improved variety of Rohu "Jayanti" developed by ICAR-CIFA (Courtesy: Rasal *et al* 2017)

Development of functional feed for sustainable Aquaculture

Functional feeds (FF) represent a transformative opportunity to enhance the profitability and sustainability of shrimp and fish aquaculture industries. These feeds play a pivotal role in increasing nutrient assimilation in aquatic animals, preventing pathogen outbreaks, and maintaining water quality in aquaculture systems. Functional feed and additives have emerged as innovative tools to enhance aquaculture production, sustainability, and profitability. These additives offer numerous physiological benefits, including increased growth rates, enhanced disease resistance, stress tolerance, and improved animal health and welfare. Probiotics, prebiotics, para-probiotics, phytochemicals, unique proteins, nucleotides, and pigments are among the key functional additives incorporated into the feed of aquatic animals. Nutritional programming, which involves administering dietary additives during specific developmental periods to exert long-lasting effects on physiology, growth, and health, is crucial in the context of functional feed additives for aquaculture. Factors such as the source and origin of additives, inclusion doses, feeding period, and fish development stage during administration influence nutritional programming.

However, it is essential to ensure appropriate formulations and incorporation rates to maximize practical benefits while minimizing anti-nutritional effects. For example, protease inhibitors can adversely affect nutrient digestibility and mineral absorption. Addressing challenges such as identifying synergistic additive combinations, developing efficient production techniques, and understanding the mechanisms of action of bioactive compounds is crucial in aquaculture.

Fig. 20.3: Functional fish feed developed by ICAR-CIFRI (under trial)

Globally, research and investment in functional fish feed development are also on the rise. Companies and research institutions in countries like Norway, China, the United States, and Thailand are at the forefront of innovation in this field, leveraging advanced technologies and scientific expertise to develop high-performance feed formulations. Additionally, collaborations and partnerships between academia, industry, and government agencies are driving progress in functional fish feed development worldwide.Collectively, these studies demonstrate the significant potential of tailored functional feed and additives to improve growth performance, enhance immune and gut health, reduce disease incidence, and ultimately increase production and profitability in aquaculture. This encompasses the latest research on the applications, outcomes, and future directions of functional feed additives for various farmed fish species, providing a comprehensive overview of current trends in the field.

Despite the availability of cheaper alternative ingredients, many of them pose challenges in terms of digestibility, water contamination, and potential pathogen proliferation. In India, as in other parts of the world, research indicates that high levels of non-digestible proteins, lipids, and complex carbohydrates in alternative ingredients can compromise the health and development of monogastric animals like fish and shrimp. The shift towards sustainable and environmentally responsible aquafeeds is driving the exploration of alternative protein and lipid sources. Future research efforts should focus on identifying novel ingredients, such as insect meal, single-cell proteins, and algae, that offer nutritional benefits while minimizing environmental impact. Additionally, research into sustainable feed production techniques, such as fermentation and biorefinery processes, can further enhance the eco-efficiency of functional feed production. This underscores the urgent need to identify alternative ingredients that are economically viable, environmentally sustainable, and safe for animal consumption.

Culture Based Fisheries (CBF) enhancement through technological interventions

Culture-based fisheries (CBF) are the fisheries practices to enhance stocks in the Inland water bodies, that don't have enough natural recruitment to sustain as a fishery. CBF practices are usually applied in small water bodies such as small reservoirs, floodplain wetlands and ox-bow lakes. Fish growth is driven by the natural productivity of the waters. Usually there is no feeding and the fish are left to forage on natural food items. India is bestowed with varied potential resources in the form of rivers and canals (1.95 lakh km); floodplain lakes (7.98 lakh hectare); ponds and tanks (24.33 lakh hectare); reservoirs (29.26 lakh hectare) and brackish water (11.55 lakh hectare). These huge resources have enormous potential for enhancing fish production, livelihood improvement and nutritional security for millions.

ICAR-CIFRI initiated the eco-friendly option of developing culture–based fisheries based on carrying capacity in small reservoirs coupled with the stock and species enhancement in medium and large reservoirs. Through successful demonstration and adoption of CIFRI technology on reservoir fisheries management, the reservoir productivity has increased substantially to 190, 98 and 34 kg/ha/year from small, medium and large reservoirs, respectively. In addition, employment of the fishermen moved upward from 30 man-days to 150 man-days/fisherman household/year through adoption of culture-based fisheries in reservoirs. Guidelines and best management practices for fisheries development in reservoirs have been formulated. Strategic plan has been developed to achieve targeted production of 50, 147 and 343 kg/ha/year from large medium and small reservoirs of the country under PMMSY towards doubling farmers income. There are many significant achievements in reservoir production across the country. The fish yield of 58 small reservoirs of Odisha has increased by 158% through adoption of culture-based fisheries.

Fish production level of 125kg/ha/yr has been achieved in selected reservoirs of Chhattisgarh, Madhya Pradesh, Jharkhand, Karnataka and Tamil Nadu through CIFRI interventions. India has vast floodplain wetland resources to the tune of 0.5 million ha situated in Ganga, Brahmaputra and Teesta-Torsa river basin. These wetlands are among the most productive resources with an average fish yield of 400kg/ha/yr against a potential of 1500-2000kg/ha/year and offers multiple ecosystem goods and services benefitting millions of people in term of livelihood and nutritional security. Research has been conducted for ecological assessments and fisheries enhancement through various interventions and management models. Most promising strategies developed for fisheries enhancement in wetlands are culture-based fisheries (CBF) and pen culture. Culture based fisheries have been adopted in wetlands of the eastern and north-eastern states of the country and have resulted in a significant increase in fish production (Fig.4).

Fig. 204: Haul of carps harvested through CBF by ICAR-CIFRI in Akaipur wetland, West Bengal

Water resources mapping for Inland fisheries development through GIS and Remote Sensing

Remote sensing offers significant potential for improving water resource management by providing valuable data on water availability, quality, and changes. As technology advances and data analysis techniques become more sophisticated, remote sensing is expected to play an increasingly crucial role in ensuring sustainable water use for future generations. Remote sensing has emerged as a powerful tool for monitoring and managing water resources due to its ability to provide accurate and timely data and its cost-effectiveness. The two main type of remote sensing data used for water resource mapping are Passive sensors: These sensors measure the natural radiation reflected from the Earth's surface, such as sunlight. Examples include multispectral sensors like Landsat and Sentinel-2, which capture data in several visible and infrared wavelengths. Active sensors: These sensors emit their own radiation and measure the reflected signal. Examples include radar sensors, which can penetrate clouds and provide data regardless of weather conditions. The application of remote sensing for water resource management offers various benefits such as accurate and timely data over large areas, cost-effective compared to traditional methods, objective and non-invasive data collection. However, it has a few limitations such as requirement of specialized skills and software for data processing and analysis, cloud cover and atmospheric conditions can affect data quality. Since 2001, ICAR-CIFRI has been at the forefront of harnessing the power of geospatial technology for the betterment of inland fisheries in India. This dedication has resulted in remarkable achievements. Water body mapping and information system by employing LISS III satellite imagery and GIS, ICAR-CIFRI meticulously mapped water bodies exceeding 10 hectares across multiple states. This comprehensive information system, developed using Microsoft Access and Visual Basic, empowers users with vital data like area, perimeter, location, and administrative Details. With financial support from the Ministry of Agriculture, ICAR-CIFRI extended water body mapping to encompass 19 states, focusing on areas exceeding 0.5 hectares. Interactive electronic maps were developed to share these critical resources with state fisheries departments, promoting transparency and accessibility. By embracing innovation and collaboration, ICAR-CIFRI continues to be a driving force in ensuring the long-term prosperity of inland fisheries and the communities that depend on them. While visual interpretation provides a valuable starting point, advanced techniques like Landsat 8 data analysis and GIS utilization offer a deeper understanding of water resources. By combining these tools with satellite data analysis, we can effectively map, monitor, and manage water resources for sustainable use and fisheries development. Hyperspectral imaging, high radiometric resolution, and frequent satellite revisit times contribute to a more detailed and dynamic understanding of water resources. When combined with water indices, these advanced techniques unlock a wealth of information, empowering us to manage our precious water resources more effectively and sustainably (Fig.5).

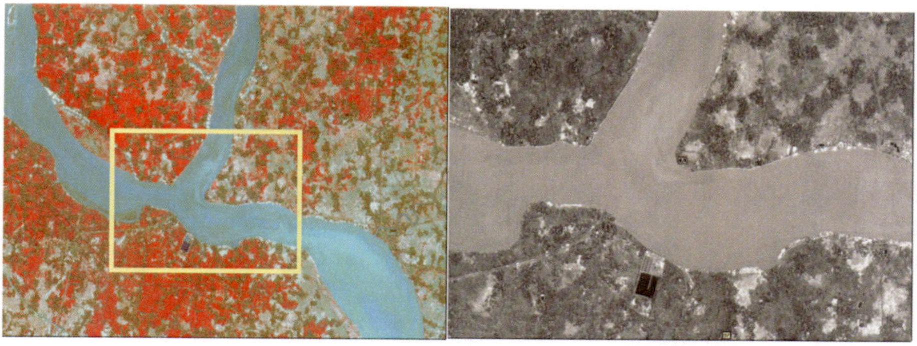

Fig. 20.5: Remote sensing image with varying spatial resolution of 24 m (left) and 6 m (right) in Ganga River

Environmental Flow for Sustenance of Riverine Ecology and Fisheries

Riverine ecosystems are facing increasing threats due to human activities like overexploitation, pollution, habitat fragmentation, and water resource development. These activities often alter the natural flow regime of rivers, leading to detrimental consequences for aquatic biodiversity, including fish populations. Maintaining natural flow patterns is crucial for the sustainability of riverine fisheries management. Environmental flow, also known as ecological flow or instream flow refers to the quantity, timing, and quality of water required to maintain the health and integrity of a river ecosystem and its dependent communities. This includes not just the minimum amount of water needed for survival, but also the seasonal fluctuations and flood events that are essential for various ecological processes. Maintaining environmental flows helps sustain the food web dynamics and over all ecosystem health and the communities that depend on them. Environmental flows, derived from ecohydrological principles, aim to mimic the natural flow patterns of a river. This includes not just the average water volume but also the seasonal fluctuations and flood events that are essential for various ecological processes. By implementing environmental flow regimes, we can create conditions that support healthy fish populations within a thriving ecosystem. Hence two distinct approaches to riverine fisheries conservation: species based and ecosystem based is the need of the hour. Species-based approaches focus on the conservation of specific fish species, often those of commercial or conservation value. This approach typically involves identifying the specific flow requirements of these target species and implementing measures to maintain suitable flow regimes. While this approach can be effective for specific fish populations, it may neglect the needs of other species and potentially disrupt the overall ecosystem balance. Ecosystem-based approaches, on the other hand, consider the entire river ecosystem as a complex interconnected system. This approach aims to maintain the diversity and health of all species, including fish, plants, insects, and microorganisms, by protecting the ecological processes that sustain them. However, there are a few challenges in implementing the environmental flow plans as it must be easy to use and cost effective, Legally robust and acceptable to stakeholders and should be holistic with due consideration to all ecosystem consequences. The biggest challenge lies in convincing policymakers of the value of environmental flows. Demonstrating the link between healthy ecosystems and human well-being is crucial for integrating environmental flows into governance structures and securing their long-term implementation. Since last decade, ICAR-CIFRI has been working on the environmental flows in different rivers including river Ramganga, Narmada, Cauvery, Teesta, Kameng and Mahanadi (Fig. 6). Major recommendations have been suggested to the respective hydropower agencies and to the respective state Departments.

Fig. 20.6: Environmental flows study in the Kameng river, Arunachal Pradesh

Conserving river Ganga fisheries and ecology through river ranching programme: A novel and recent approach

The river Ganga is the country's national and most revered river rises from the Gangotri Glacier, running approximately 2,525 kilometers. The 'Namami Gange' program envisages a holistic approach towards rejuvenation, conservation, and cleaning of the river Ganga. The river supports a total of 198 fish species (189 native and 9 exotic) belonging to 143 genera, 65 families, and 55 orders from upper to the lower stretch of the river as per the recent reports of ICAR- CIFRI, Barrackpore. Since the river has been infiltrated with alien/ exotic species from past few decades, utmost importance to restore these viable native populations along with all endemic and endangered biodiversity of the river is highly needed. Restoration of aquatic life in terms of fishes is vital for harnessing biological diversity of the river and will ultimately indicate about level of cleanliness. Ranching can simply be defined as the process through which economically important fishes are stocked in river for future harvest. In this context, ICAR- Central Inland Fisheries Research Institute under the project National Mission for Clean Ganga (NMCG) initiated by Ministry of Water Resources, River Development and Ganga Rejuvenation is primarily associated to practice river ranching of few valued fish species like Rohu, Catla, Mrigal, Calbasu collectively known as Indian Major Carps and Cold-water fishes like Mahseer in upper stretches of the river. River ranching, a form of ex-situ conservation, is a method of conservation in reviving the fish stocks maintaining the genetically diverse sustainable fisheries. This mega approach is a first kind in India initiated by ICAR- CIFRI as a part of conserving and reviving the native Gangetic fish germplasm. The sites for ranching along the river Ganga were determined based on species distribution and commercial Catch Per Unit Effort (CPUE) data generated by CIFRI. The institute (CIFRI) have ranched a total of 94.53 lakhs of indigenous fishes comprising of Indian Major Carps, Mahseer and catfishes in 100 different events over the period of 2017-2023 under the esteemed Namami Gange Project (Fig. 7). The ranching programmes are carried out with fish size range of 10-15cm with average weight of 100-120g. The fishes are bred artificially using wild brooders of river Ganga and finally are grown up to fingerling/advanced fingerling size before releasing them in the river. The fish ranching programmes were performed in different states along river Ganga at various sites. The maximum ranching programme was carried out in the state of West Bengal (55%), followed by Uttar Pradesh (28%), Bihar (9%), Jharkhand (7%) and Uttarakhand (1%) during the period of 2017-2023. Apart of from ranching of Indian Major Carps, CIFRI is also ranching prized Hilsa towards restoration of the fish species in the upstream of Farakka barrage in West Bengal. A total of 12.85 lakhs of fertilized hilsa eggs and more than 8,500 spawns were also ranched at both upstream and downstream of Farakka barrage in the Ganga River during the period 2020-2023. In addition to this, a total of 2,568 number of adult hilsa was tagged and released in river Ganga to understand the migratory pattern of the species. The impact of river ranching has been assessed by ICAR-CIFRI from the year 2020-23. Landing of Rohu, Catla and Mrigal during 2018-2022 has shown an increasing trend in Ganga. However, there was a lesser catch in 2021 due to the lockdown due to the Covid-19 pandemic. A significant impact of river ranching was observed when the present landing data (2023) from Prayagraj, Uttar Pradesh was compared with the time series data of ICAR-CIFRI records (from 1960 onwards) showing that the landing of IMC increased by 24.70% when compared to 1959 (11.70 tonnes). Similar results were also achieved from sites like Varanasi (increase IMC landing of 41.03%) and Buxar, Bihar. This is the novel concept where the brooders are collected from the same system and the seeds are released to the depleted stretches of the same ecosystem i.e river in order to avoid the genetic contamination. This protocol not only avoid genetic contamination but also improve the depleted fish stock in its natural habitat.

Fig. 20.7: River ranching of Indian Major Carps in river Ganga at Farakka, West Bengal

Valuation of Ecosystem Services, Governance and Tenure in inland open waters

India's rich inland open water resources, including rivers, wetlands, and reservoirs, play a vital role in the country's ecological balance, food security, and livelihoods of millions. However, these precious resources face increasing pressure from various stakeholders, leading to overexploitation, degradation, and conflicts. Economic valuation offers a powerful tool to understand the true worth of these resources and guide sustainable management practices. Traditional cost-benefit analysis often overlooks the environmental benefits of inland open water ecosystems, resulting in biased decisions favoring development over conservation. Economic valuation provides a framework to quantify and compare the various goods and services provided by these ecosystems. The direct benefits of these ecosystem services comprise of fisheries, agriculture, water supply, recreation, and tourism. However, many indirect benefits can be derived from these open water resources such as flood control, nutrient cycling, water purification, and habitat for biodiversity. The evaluation of ecosystem services also aims to make use of non-use benefits by considering its existence value, aesthetic value, and cultural heritage. By assigning monetary values to these benefits, we can compare different management options We can ensure the optimal resource allocation by evaluating the trade-offs between development and conservation. It makes easy to demonstrate the economic value of preserving these ecosystems, garnering support from policymakers and stakeholders and there by justifying the conservation efforts. Valuation of services also promote in designing effective policies that discourage environmentally harmful practices. Some environmental benefits, like biodiversity and cultural heritage, are difficult to quantify in monetary terms. Additionally, valuation studies can be complex and require specialized expertise. However, despite these limitations, economic valuation remains a valuable tool for guiding sustainable management decisions. The scope of economic valuation techniques in inland open water ecosystems includes market valuation by estimating the market prices of goods and services derived from the ecosystem, such as fish catches or recreational fees and also enable to analyze the travel expenses incurred by people visiting the ecosystem for recreation or tourism. The contigent valuation method can be applied by surveying individuals about their willingness to pay for the preservation of the ecosystem or its benefits. Effective management of inland open water resources requires not only economic valuation but also strong governance structures. This includes secure tenure and access rights by ensuring that local communities have access and control over the resources, fostering responsible stewardship. Involving stakeholders in decision-making processes, promoting transparency and accountability in a collaborative mode is highly needed. Thus, it can be concluded that economic valuation offers a powerful tool to translate the intrinsic value of inland open water resources into a language understood by policymakers and stakeholders. By integrating valuation with strong governance practices, we can ensure the sustainable management of these vital ecosystems for the benefit of present and future generations.

Application of Artificial Intelligence towards sustainable fisheries in Inland Open waters

The application of Artificial Intelligence (AI) is rapidly transforming various sectors, and the fisheries industry is no exception. Inland open waters, including rivers, floodplains, wetlands, and reservoirs, contribute significantly to global biodiversity and provide vital ecosystem services. Advancements in AI and related technologies, such as the Internet of Things (IoT) and machine learning, offer promising solutions to address these challenges and improve the sustainability and efficiency of inland open water fisheries. The application of AI in aquaculture includes Environment monitoring with the help of drones equipped with sensors can collect real-time data on water quality parameters like dissolved oxygen, temperature, and turbidity, allowing farmers to optimize fish growth conditions (Fig. 8). AI-powered feeders analyze fish behavior and appetite, dispensing precise amounts of feed at the right time, minimizing waste and maximizing efficiency. Machine learning algorithms can analyze fish behavior, appearance, and genetic data to detect disease outbreaks early, enabling preventive measures and reducing mortality. Vision-based AI systems can automatically track and analyze fish swimming patterns, size, and behavior, identifying potential health issues and improving overall stock health. AI algorithms can identify and remove unhealthy or deformed fish fry from rearing tanks, ensuring stronger fish populations. AI models can predict optimal harvest times based on environmental conditions, fish growth, and market demand, maximizing yield and profitability. However, for the implementation of AI technology requires significant investment and expertise, posing challenges for smaller farms and developing countries. Effective AI algorithms rely on accurate and comprehensive data, which may not be readily available in all regions or situations. Transparency and responsible use of AI data are crucial to avoid ethical dilemmas and ensure fair practices within the industry. Despite the challenges, AI holds immense potential for revolutionizing fisheries management. By optimizing practices, improving data-driven decision-making, and enhancing sustainability, AI can contribute significantly to food security and a thriving future for the fisheries sector. As research and development continue, AI applications will become more accessible and adaptable, paving the way for a more efficient and resilient fisheries industry.

Fig. 20.8: Use of drone by ICAR-CIFRI for water sampling in river Ganga

Field Exposure and River ranching programme in river Ganga

The MGCC trainees participated actively in the field exposure visit and hands on training including the river ranching programme in river Ganga (Fig. 9-14). This helped the trainees to learn the concept and application in the river Mekong towards conservation of depleted fish species. In addition, exposure visit to the carp farms helped them to learn the breeding and possible adaptation of Indian carps in the cage culture.

Fig. 20.9: Association of MGCC trainees in Gangetic fish conservation programme

Fig. 20.10: Visit of MGCC trainees to the Carp hatchery and nursery ponds

Fig. 20.11: River ranching programme by MGCC trainees in river Ganga

Fig 20.12: Field exposure visit of MGCC trainees

Fig. 20.13: Field exposure visit of MGCC trainees

Fig. 14: Field exposure visit of MGCC trainees

Epilogue

The Mekong-Ganga Cooperation (MGC) represented by five ASEAN countries, namely, Cambodia, Lao PDR, Myanmar, Thailand, and Vietnam was initially called Ganga Suvarnabhumi Programme (GMSP). Both the Ganga and the Mekong are civilizational rivers, and the MGC initiative aims to facilitate closer contacts among the people inhabiting these two major river basins. In both the river systems, fisheries play a significant role on the livelihood of the millions. However, increased anthropogenic activities coupled with climate change have posed serious threats to the fisheries and river ecology. With this backdrop, realizing the importance, a training programme on sustainable fisheries for these countries was organised at ICAR-CIFRI, Barrackpore to share the fisheries research and development programmes and activities in these two basins. The major approach was through a) trainee central approach: the overall training sessions were designed to facilitate maximum experience sharing and interaction among participants and resource persons b) focus on skill enhancement/improvement: the analytical assessment and application tools were developed and provided to participants for planning, operation, monitoring, and evaluation of management practices along with theoretical and practical sessions c) focus on practical and problem-solving inclination: trainees were encouraged to present their cases, identify the issues, and problems and apply the knowledge and skills to plan for appropriate management practices/solutions with proposed approaches. The post-training evaluation showed high satisfaction and brough important areas those need for further collaborations. Major areas including selective breeding, culture-based fisheries, environmental flows, and river ranching programmes were felt to be important for both Mekong and Ganga River basins towards improving the livelihood and achieving sustainable development goal for the dependents of both the river basins. Fisheries enhancement through technological interventions/management strategies and conservation of inland open water fisheries resources are essential for ensuring sustainable fish production, livelihood and nutritional security. This training helped to learn the advances in fisheries management, existing practices and technological advancements for sustainable fisheries from vast inland waters. By implementing the proposed strategies and activities, we can enhance the sustainability, resilience and fish yield from open waters while safeguarding resources for future generations.

Acknowledgements

The authors are greatly acknowledged to the Ministry of External Affairs for the financial supports and special thanks to Ms. Paramita Tripathi Joint Secretary, Shri Harsh Kumar Chirania Under Secretary and Mr. Puran Patwari, Consultant, Indo-Pacific Division Ministry of External Affairs for the supports. The authors are thankful to Dr. Dr. Himanshu Pathak Secretary, DARE & Director General, Dr. J. K. Jena Deputy Director General (Fisheries Science), Dr. B.P Mohanty Assistant Director General (Inland Fisheries), Indian Council of Agricultural Research, Ministry of Agriculture and Farmers Welfare, New Delhi for the administrative supports. The authors also thank all the resource experts, administrative staff, research scholars and technical staff of ICAR-CIFRI for the support during the programme.

References

Chircop, A., 1998. Introduction to capacity-building section. Ocean and Coastal Management, 38(1), pp.67-68.

Espinoza-Tenorio, A., Espejel, I. and Wolff, M., 2011. Capacity building to achieve sustainable fisheries management in Mexico. Ocean & coastal management, 54(10), pp.731-741.

FAO. 2022. The State of World Fisheries and Aquaculture 2022. Towards Blue Transformation. Rome, FAO. https://doi.org/10.4060/cc0461en

UNCED, 1992. Agenda 21 on United Nations Conference on Environment and Development, Rio de Janerio, Brazil, 3 to 14 June 1992. Pp. 351. https:// efaidnbmnnnibpcajpcglclefindmkaj/ https://sustainabledevelopment.un.org/content/documents/Agenda21.pdf

21

Capacity Building in Nutrition Sensitive Agriculture for Addressing Global Malnutrition

Pragati Shukla and Veenita Kumari

National Institute of Agricultural Extension Management, Hyderabad

To combat malnutrition and micronutrient deficiencies, nutrition-sensitive agriculture is a food-based strategy to agricultural development that emphasizes dietary diversity, fortification, and consumption of nutrient-rich foods. This strategy recognizes the importance of consuming a wide range of foods for optimal nutrition and acknowledges the social and economic significance of the food and agriculture industries in supporting rural livelihoods (WHO, 2014)

The overarching goal of nutrition-sensitive agriculture is to enhance the ability of the global food system to produce nutritious outcomes. An increasing number of governments, donor agencies, and development organizations are endorsing nutrition-sensitive agriculture as a means to achieve their developmental objectives. However, relying solely on nutrition-specific interventions falls short of meeting global targets for improving nutrition. Agriculture, with its significant potential to influence nutrition outcomes, must also play a role.

Transforming agriculture into a more nutrition-sensitive sector requires a fresh approach to thinking, planning, implementing, and collaborating. It necessitates the active involvement of diverse stakeholders from various sectors and the identification of opportunities to integrate nutrition objectives into agro-food systems. Initial steps include designing conceptual frameworks that identify the multiple pathways through which agriculture can impact nutrition. Several conceptual frameworks have been developed, highlighting the dynamic and multifaceted linkages between agriculture, health, and nutrition.

The significance of capacity building in nutrition-sensitive agriculture for addressing global malnutrition is profound and multi-faceted, making it a crucial subject. Here are several key points highlighting its importance:

Holistic Approach to Malnutrition: Nutrition-sensitive agriculture takes a comprehensive approach to addressing malnutrition by focusing on the entire food system, from production to consumption. By integrating nutrition objectives into agricultural practices, offers a holistic solution to the complex problem of malnutrition, which encompasses both undernutrition and overnutrition.

Sustainable Development Goals (SDGs): Nutrition-sensitive agriculture aligns with several of the United Nations Sustainable Development Goals (SDGs), including SDG 2 (Zero Hunger) and SDG 3 (Good Health and Well-being). As countries strive to achieve these goals by 2030, capacity building in nutrition-sensitive agriculture becomes essential for implementing effective strategies that can contribute to global efforts to eradicate hunger and improve nutrition.

Empowerment of Vulnerable Groups: Nutrition-sensitive agriculture emphasizes the importance of empowering vulnerable groups, such as women and smallholder farmers, who are disproportionately affected by malnutrition. Capacity-building initiatives provide these groups with the knowledge, skills, and resources they need to adopt sustainable agricultural practices, improve food security, and enhance their nutritional well-being.

Resilience to Climate Change: Climate change poses significant challenges to food security and nutrition worldwide. Nutrition-sensitive agriculture promotes climate-smart agricultural practices that enhance resilience to climate change while also improving nutrition outcomes. Capacity-building efforts equip farmers and agricultural extension workers with the tools and techniques necessary to adapt to changing environmental conditions and mitigate the impacts of climate variability on food production and nutrition.

Cross-Sectoral Collaboration: Addressing malnutrition requires collaboration across multiple sectors, including agriculture, health, education, and social welfare. Capacity building initiatives in nutrition-sensitive agriculture facilitate cross-sectoral collaboration by bringing together stakeholders from diverse backgrounds to develop integrated solutions that address the underlying drivers of malnutrition and promote sustainable development.

Global Health Security: Malnutrition not only has significant implications for individual health and well-being but also poses broader risks to global health security. Poor nutrition weakens immune systems, increases susceptibility to infectious diseases, and contributes to the spread of foodborne illnesses. By improving nutrition outcomes through nutrition-sensitive agriculture, capacity-building efforts can help strengthen global health security and mitigate the risk of pandemics and other health crises.

In summary, capacity building in nutrition-sensitive agriculture is essential for addressing global malnutrition comprehensively and sustainably. By empowering individuals and communities to adopt nutrition-sensitive agricultural practices, capacity building initiatives can contribute to achieving the SDGs, empowering vulnerable groups, building resilience to climate change, fostering cross-sectoral collaboration, and enhancing global health security. These themes highlight the critical significance of the topic for the content of a book chapter on addressing global malnutrition.

Identified pathways through which agricultural interventions can impact nutrition include:

1. Food access from own-production.
2. Income from the sale of commodities produced.
3. Food prices from changes in supply and demand.
4. Women's social status and empowerment through increased access to resources.
5. Women's time through participation in agriculture.
6. Women's health and nutrition through engagement in agriculture.

Food access from own-production

This refers to the direct access individuals or households have to food produced through their own agricultural activities. When farmers grow crops or raise livestock for their own consumption, it ensures a more reliable and immediate source of food. This can contribute to improved nutrition by providing access to fresh, nutritious foods without relying solely on external markets.

Income from the sale of commodities produced

Agricultural production often serves as a source of income for farmers. By selling their produce in local markets or to larger distributors, farmers earn income that can be used to purchase additional food, as well as other necessities such as healthcare, education, and improved housing. Increased income from agricultural sales can enhance household food security and access to a diverse range of foods, ultimately improving nutrition outcomes.

Food prices from changes in supply and demand

Changes in agricultural supply and demand can have significant effects on food prices, which in turn influence consumers' purchasing power and food access.

For example, a decrease in crop yields due to adverse weather conditions may lead to higher food prices, making it more difficult for low-income households to afford nutritious foods. Understanding these dynamics is important for policymakers and stakeholders to implement interventions that stabilize food prices and ensure affordability for all consumers.

Women's social status and empowerment through increased access to resources

Women play a crucial role in agricultural production and food security, particularly in many rural communities around the world. Empowering women by providing them with equal access to resources such as land, credit, and agricultural inputs can have profound effects on household nutrition. When women have greater control over resources and decision-making within the household, they are more likely to prioritize the nutritional needs of their families, leading to improved dietary diversity and overall nutrition outcomes.

Women's time through participation in agriculture

Women often bear the primary responsibility for agricultural labor, including planting, harvesting, and processing crops, as well as caring for livestock and managing household gardens. The time women spend engaged in agricultural activities can impact their ability to engage in other activities, including childcare, education, and income-generating opportunities. Balancing these responsibilities is crucial for women's well-being and the overall health of their families.

Women's health and nutrition through engagement in agriculture

Women's involvement in agriculture can have both positive and negative effects on their health and nutrition. On one hand, access to nutritious foods produced through agriculture can improve women's dietary intake and overall health. On the other hand, the physical demands of agricultural labor, combined with limited access to healthcare and nutrition services, can increase women's risk of malnutrition and other health problems. Promoting women's health and well-being within agricultural contexts requires comprehensive interventions that address both nutritional and healthcare need (IFPRI, 2011).

Recognizing the interconnectedness of agriculture and nutrition, as well as the mediating role of women's status and empowerment, is crucial for developing programs and investments aimed at using agriculture to enhance nutrition.

Objectives

1. Develop a conceptual understanding of the concept, need, and approaches of nutrition-sensitive agriculture.
2. Understand food-based systems globally and their potential to address malnutrition.
3. Expose participants to different approaches and models of nutrition-sensitive agriculture.
4. Develop an action plan to operationalize the concepts learned during the training program.

Strategies for implementation of NSA Capacity Building Program

Capacity building in nutrition-sensitive agriculture for addressing global malnutrition involves a range of strategies and approaches aimed at equipping individuals, communities, and institutions with the knowledge, skills, and resources needed to promote nutrition-sensitive practices within the agricultural sector. Here are some key ways in which capacity building can be done:

1. **Training Programs and Workshops**: Organize training programs and workshops targeted at farmers, agricultural extension workers, policymakers, researchers, and other relevant stakeholders. These programs should cover topics such as sustainable agriculture, dietary diversity, food fortification, post-harvest management, and value addition to agricultural products.
2. **Curriculum Development**: Integrate nutrition-sensitive agriculture into educational curricula at agricultural universities, colleges, and vocational training institutes. Develop specialized courses, modules, and training materials that incorporate principles of nutrition-sensitive agriculture and provide hands-on learning experiences for students and practitioners.
3. **Demonstration Farms and Field Days**: Establish demonstration farms and organize field days to showcase best practices in nutrition-sensitive agriculture. These practical demonstrations allow farmers to see firsthand how adopting nutrition-sensitive practices can improve crop yields, enhance soil health, conserve natural resources, and promote dietary diversity.
4. **Knowledge Sharing Platforms**: Create online and offline knowledge sharing platforms, such as websites, mobile applications, community radio programs, and farmer field schools, to disseminate information

and resources on nutrition-sensitive agriculture. Encourage peer-to-peer learning and knowledge exchange among farmers and agricultural professionals.

5. **Partnership and Collaboration**: Foster partnerships and collaboration among government agencies, non-governmental organizations (NGOs), research institutions, international organizations, and private sector stakeholders involved in agriculture, nutrition, and rural development. Pool resources, expertise, and networks to support capacity building initiatives and scale up successful interventions.
6. **Policy Advocacy and Institutional Strengthening:** Advocate for supportive policies, regulations, and investments that promote nutrition-sensitive agriculture and create an enabling environment for capacity building activities. Strengthen institutional capacities within government agencies, extension services, and farmer organizations to mainstream nutrition-sensitive approaches into agricultural development programs and projects.
7. **Monitoring and Evaluation**: Establish robust monitoring and evaluation mechanisms to track the progress and impact of capacity building efforts in nutrition-sensitive agriculture. Collect data on indicators such as adoption of nutrition-sensitive practices, changes in dietary patterns, household food security, and nutritional outcomes to assess the effectiveness of interventions and inform future programming.
8. **Community Engagement and Empowerment**: Engage local communities in the design, implementation, and evaluation of capacity building initiatives to ensure relevance, ownership, and sustainability. Empower farmers, especially women and marginalized groups, to actively participate in decision-making processes and become agents of change in promoting nutrition-sensitive agriculture at the grassroots level.

By employing these approaches in a coordinated and participatory manner, capacity building in nutrition-sensitive agriculture can contribute to addressing global malnutrition and advancing sustainable development goals related to food security, nutrition, health, and poverty alleviation.

Several strategies can be particularly effective in capacity building for nutrition-sensitive agriculture to address global malnutrition comprehensively. Here are some key approaches:

i. **Multi-Stakeholder Collaboration**: Foster collaboration among governments, NGOs, academia, research institutions, international organizations, and the private sector. This collaboration can facilitate the sharing of expertise, resources, and best practices, leading to more comprehensive and impactful capacity building initiatives.

ii. **Tailored Training Programs**: Develop tailored training programs that cater to the specific needs and contexts of different target groups, such as farmers, extension workers, policymakers, and healthcare professionals. These programs should incorporate hands-on learning experiences, case studies, and practical demonstrations to enhance understanding and retention of key concepts.

iii. **Integration into Existing Systems**: Integrate capacity building for nutrition-sensitive agriculture into existing agricultural extension services, educational curricula, and rural development programs. This integration ensures sustainability and scalability by leveraging existing infrastructure and networks.

iv. **Gender-Responsive Approaches**: Adopt gender-responsive approaches that promote women's empowerment and participation in agriculture. Provide training and resources specifically targeted at women farmers to address gender disparities in access to land, credit, inputs, and decision-making power.

v. **Innovative Technologies**: Harness the potential of innovative technologies, such as mobile applications, e-learning platforms, and remote sensing tools, to deliver training and extension services to remote and underserved communities. These technologies can overcome geographical barriers and reach a wider audience more cost-effectively.

vi. **Community-Led Initiatives**: Empower local communities to take ownership of capacity building initiatives and tailor interventions to their specific needs and priorities. Engage community leaders, farmer groups, and grassroots organizations in the planning, implementation, and evaluation of programs to ensure relevance and sustainability.

vii. **Policy Advocacy**: Advocate for supportive policies, incentives, and regulations that promote nutrition-sensitive agriculture and create an enabling environment for capacity building efforts. Engage policymakers at the local, national, and international levels to prioritize investments in agriculture, nutrition, and rural development.

viii. **Monitoring and Evaluation**: Establish robust monitoring and evaluation systems to track the progress and impact of capacity building activities. Collect data on key indicators, such as adoption of nutrition-sensitive practices, dietary diversity, food security, and nutritional outcomes, to measure effectiveness and inform evidence-based decision-making.

By combining these strategies in a coordinated and context-specific manner, capacity building for nutrition-sensitive agriculture can play a crucial role in addressing global malnutrition and contributing to sustainable development goals related to food security, nutrition, health, and poverty reduction.

Areas in which capacity building is required to address global malnutrition

To effectively address global malnutrition, capacity-building efforts should target various key areas across different sectors. Here are some of the areas where capacity building would be required:

Agricultural Practices: Capacity building in sustainable and nutrition-sensitive agricultural practices is essential to improve crop yields, promote dietary diversity, and enhance the nutritional value of food. Training programs should focus on agroecology, soil management, water conservation, integrated pest management, and climate-smart agriculture.

Nutrition Education: Capacity building in nutrition education is necessary to raise awareness about the importance of balanced diets, micronutrient-rich foods, and healthy eating habits. Training should cover topics such as food groups, dietary guidelines, meal planning, food safety, and the prevention of malnutrition-related diseases.

Food Processing and Preservation: Capacity building in food processing and preservation techniques is crucial to minimize post-harvest losses, improve food storage, and increase the availability of nutritious foods year-round. Training programs should include methods for drying, canning, fermentation, and other food preservation methods.

Healthcare Services: Capacity building in healthcare services is essential to provide early detection, treatment, and management of malnutrition-related conditions. Training should focus on screening tools, diagnostic techniques, therapeutic interventions, and counseling approaches for healthcare providers.

Community Empowerment: Capacity building in community empowerment strategies is necessary to engage local communities in identifying, prioritizing, and implementing solutions to malnutrition. Training should include participatory approaches, community mobilization techniques, and advocacy skills to empower communities to take ownership of their health and nutrition outcomes.

Policy Development: Capacity building in policy development and advocacy is critical to create an enabling environment for addressing malnutrition at the national and international levels. Training should focus on evidence-based policymaking, policy analysis, stakeholder engagement, and communication strategies to advocate for nutrition-sensitive policies and programs.

Research and Innovation: Capacity building in research and innovation is necessary to generate new knowledge, technologies, and solutions for addressing malnutrition effectively. Training should cover research methodologies, data analysis techniques, and dissemination strategies to ensure that research findings translate into actionable recommendations for policymakers and practitioners.

Cross-Sectoral Collaboration: Capacity building in cross-sectoral collaboration is essential to foster partnerships and coordination among different stakeholders working in agriculture, health, education, social welfare, and other relevant sectors. Training should promote interdisciplinary approaches, networking opportunities, and shared learning platforms to facilitate collaboration and maximize impact.

By investing in capacity building across these key areas, stakeholders can strengthen their ability to address the complex and multifaceted challenges of global malnutrition and work towards sustainable solutions that improve the health and well-being of populations worldwide.

Roles and Responsibilities of Institutions Towards Capacity Building on NSA

Institutions play a crucial role in capacity building for nutrition-sensitive agriculture (NSA) by providing leadership, resources, expertise, and coordination to support effective training and development initiatives. Here are some of the key responsibilities of institutions in this regard:

Curriculum Development: Institutions should develop and update curricula, training materials, and educational resources that incorporate principles of NSA. This includes designing courses, modules, and workshops that cover topics such as sustainable agriculture, food security, nutrition education, and cross-sectoral collaboration.

Training Delivery: Institutions should organize and facilitate training programs, workshops, seminars, and conferences on NSA for a diverse range of stakeholders, including farmers, agricultural extension workers, policymakers, researchers, healthcare providers, and community leaders. Training should be delivered through various formats, such as in-person sessions, online courses, and blended learning approaches.

Capacity Building Initiatives: Institutions should design and implement capacity building initiatives that address the specific needs and priorities of target groups and communities. This may involve conducting needs assessments, identifying training gaps, and tailoring interventions to enhance skills, knowledge, and competencies related to NSA.

Partnership and Collaboration: Institutions should foster partnerships and collaboration with government agencies, NGOs, research institutions, international organizations, and the private sector to leverage resources, expertise, and networks for capacity building initiatives. Collaboration can enhance the reach, impact, and sustainability of training programs and facilitate the exchange of best practices and lessons learned.

Research and Innovation: Institutions should conduct research and innovation activities to generate new knowledge, technologies, and solutions for addressing malnutrition through NSA. This may involve conducting applied research, field trials, pilot projects, and case studies to evaluate the effectiveness of different NSA interventions and inform evidence-based decision-making.

Policy Advocacy: Institutions should advocate for supportive policies, regulations, and investments that promote NSA and create an enabling environment for capacity building efforts. This includes engaging policymakers, advocating for funding and resources, and participating in policy dialogue and decision-making processes at the local, national, and international levels.

Monitoring and Evaluation: Institutions should establish monitoring and evaluation systems to assess the impact and effectiveness of capacity building initiatives on NSA. This includes collecting data on key indicators, measuring outcomes and outputs, and conducting periodic reviews and assessments to track progress, identify challenges, and inform continuous improvement.

Knowledge Sharing and Dissemination: Institutions should disseminate information, resources, and best practices on NSA through various knowledge sharing platforms, such as websites, publications, newsletters, and social media channels. This helps to raise awareness, build networks, and facilitate peer-to-peer learning among stakeholders involved in NSA capacity building efforts.

By fulfilling these responsibilities, institutions can contribute to strengthening the capacity of individuals, communities, and organizations to promote NSA and address global malnutrition comprehensively and sustainably.

Key Recommendations for Policy makers and implementers

Policy makers and implementers play a critical role in shaping the environment and resources available for capacity building in nutrition-sensitive agriculture (NSA). Here are some key recommendations for them:

For Policy Makers

1. Integrate Nutrition-Sensitive Agriculture into Policy Frameworks: Ensure that national agricultural and nutrition policies explicitly incorporate goals, strategies, and resources for promoting NSA. This integration can help mainstream nutrition considerations across the agricultural sector and ensure coherence with broader development objectives.
2. Allocate Adequate Resources: Prioritize funding and resource allocation for capacity building initiatives in NSA, including training programs, research activities, extension services, and infrastructure development. Allocate sufficient budgets and mobilize additional resources from domestic and international sources to support these efforts.
3. Promote Cross-Sectoral Collaboration: Foster collaboration and coordination among relevant ministries, agencies, and stakeholders involved in agriculture, health, education, social welfare, and other sectors. Establish multi-sectoral platforms, task forces, and working groups to facilitate information sharing, policy dialogue, and joint planning for NSA.
4. Incentivize Adoption of Nutrition-Sensitive Practices: Develop policies, incentives, and subsidies to encourage farmers and agricultural producers to adopt nutrition-sensitive practices, such as diversifying crops, improving soil fertility, reducing post-harvest losses, and promoting agroecological approaches. Provide targeted support to smallholder farmers, women, and marginalized groups.
5. Strengthen Monitoring and Evaluation Systems: Establish robust monitoring and evaluation systems to track progress, measure impact, and ensure accountability for NSA initiatives. Collect data on key indicators, such as dietary diversity, food security, nutritional outcomes, and adoption of nutrition-sensitive practices, to inform evidence-based decision-making.
6. Invest in Research and Innovation: Support research and innovation in NSA to generate new knowledge, technologies, and solutions for

addressing malnutrition effectively. Prioritize applied research, field trials, and pilot projects that evaluate the effectiveness of different NSA interventions and identify best practices for scaling up.

7. Build Institutional Capacity: Strengthen institutional capacities within government agencies, research institutions, extension services, and farmer organizations to implement NSA policies and programs effectively. Provide training, technical assistance, and support for staff development to enhance skills, knowledge, and competencies related to NSA.

For Implementers

Tailor Interventions to Local Contexts: Adapt capacity building interventions to the specific needs, priorities, and capacities of target groups and communities. Consider local agroecological conditions, cultural practices, and socioeconomic dynamics when designing and delivering training programs, extension services, and outreach activities.

1. Promote Participation and Ownership: Engage stakeholders, including farmers, women, youth, and community leaders, in the planning, implementation, and evaluation of capacity building initiatives. Foster participatory approaches, community mobilization, and empowerment strategies to ensure that interventions are inclusive, relevant, and sustainable.

2. Facilitate Knowledge Sharing and Learning: Create opportunities for peer-to-peer learning, knowledge exchange, and networking among stakeholders involved in NSA capacity building efforts. Establish learning platforms, farmer field schools, and demonstration sites to showcase best practices, share experiences, and build technical skills.

3. Provide Access to Resources and Inputs: Ensure that farmers have access to essential resources, inputs, and support services needed to adopt nutrition-sensitive practices effectively. This may include seeds, fertilizers, pesticides, irrigation equipment, credit, market linkages, and extension advice tailored to their specific needs and circumstances.

4. Strengthen Extension Services: Strengthen agricultural extension services to provide timely, relevant, and gender-sensitive advice and support to farmers on NSA. Train extension workers, equip them with up-to-date information and tools, and incentivize their performance to enhance their effectiveness in promoting nutrition-sensitive practices.

5. Monitor and Evaluate Impact: Monitor progress, measure impact, and evaluate outcomes of capacity building interventions to assess their

effectiveness and identify areas for improvement. Collect feedback from participants, gather data on adoption rates and behavioral changes, and use evidence to inform programmatic decisions and adjust strategies as needed.

6. Advocate for supportive policies, regulations, and investments that create an enabling environment for nutrition-sensitive agriculture (NSA) and capacity building initiatives. This includes advocating for policies that incentivize the adoption of NSA practices, such as crop diversification, sustainable land management, and conservation agriculture. Additionally, advocate for regulations that promote food safety, quality standards, and fair trade practices to ensure that nutritious foods are accessible and affordable for all. Moreover, lobby for investments in agricultural research, extension services, rural infrastructure, and social protection programs to strengthen the resilience and sustainability of food systems and improve nutrition outcomes. By advocating for supportive policies and investments, we can create the necessary conditions for the NSA to thrive and contribute to ending global malnutrition.

MANAGE Initiative Towards Capacity Building of Extension Personnel on Nutrition-Sensitive Agriculture and trained 410 officials on Nutrition-Sensitive Agriculture. These officials are from the ICAR system, SAUs, CAUs, KVKs, and NGOs. These training programs were conducted through online and offline mode. The topics covered are: Indicators of Food & Nutritional Security, Nutrition Sensitive Agriculture- An Overview, Urban Agriculture and Nutrition, Indicators for Nutrition-Sensitive Agriculture, Addressing Gender in Nutrition Sensitive Agriculture, Involving Men in Nutrition, Nutritional Situation in the country- Global perspective, Nutrition Sensitive Value Chains, Nutrition Sensitive Post-harvest handling, Storage and Processing, Food Loss and Waste: Prevention, Reduction and Management, Promoting Nutrition Sensitive Agriculture in Extension Advisory Services, Nutrition Sensitive Livestock Practices, Nutrition Education and Behaviour change communication, Community Engagement: Strategies for engaging communities in NSA, Emerging Technologies in Agriculture, Food Based Interventions to Address Malnutrition, Monitoring and Evaluation of Nutrition Sensitive Agriculture Programs, Achieving Zero Hunger- Measures and Strategies, Indigenous Knowledge and Traditional Agriculture, Climate-Smart Agriculture Practices for Better Nutrition, Nutrition Sensitive Fisheries Practices, Assessment of Food Consumption Pattern and Dietary Needs, Nutrition Sensitive Horticultural Practices, Global Governance and Partnerships, Climate Change and its implications on Agriculture & Food Security, Diversification and sustainable intensification of agricultural production.

Centre for Gender in Agriculture, Food & Nutritional Security & Urban Farming, MANAGE organized a 14-day International ITEC program on "Nutrition Sensitive Agriculture" for international delegates from 22 countries, held from March 6 to March 19, 2024, sponsored by the Ministry of External Affairs, Government of India.

Glimpse of Training Programme

Fig. 21.1: Session on Nutrition Education and Behaviour change communication

Fig. 21.2: Session on Involvement of Men in Nutrition

Fig. 21.3: Session on Food Based Interventions to Address Malnutrition

Fig. 21.4: Visit to Haribabu Natural Farming Maheshwaram Village, Mahboobnagar

Fig. 21.5: Visit to Dangoria Charitable Trust, Narsapur

Fig. 21.6: Visit to National Institute of Nutrition

Outcomes from the course

- Enhanced knowledge and understanding among the participants on Nutrition-Sensitive Agriculture from an Indian Context.
- Capable of implementing Nutrition sensitive approaches and models in their respective work areas
- Well-informed and equipped to handle community malnutrition as EAS/RAS Providers.

Conclusion

In conclusion, capacity building in nutrition-sensitive agriculture (NSA) stands as a pivotal strategy in our collective efforts to combat global malnutrition comprehensively and sustainably. By equipping individuals, communities, and institutions with the knowledge, skills, and resources needed to promote NSA practices, we can address the complex and interconnected challenges of food insecurity, dietary deficiencies, and poor nutrition outcomes.

Through this chapter, we have explored the significance of NSA in addressing global malnutrition, the key areas where capacity building is required, and the respective roles and responsibilities of institutions, policymakers, and implementers. We have underscored the importance of integrating nutrition considerations into agricultural policies, fostering cross-sectoral collaboration, and prioritizing investments in research, training, and infrastructure.

Moving forward, it is imperative that we continue to prioritize and scale up capacity building efforts in NSA, leveraging innovative approaches, partnerships, and technologies to maximize impact and reach. By investing in human capital, strengthening institutional capacities, and advocating for supportive policies, we can create an enabling environment for sustainable agricultural development and improved nutrition outcomes for all.

Ultimately, the success of our endeavors in NSA capacity building will be measured by our ability to empower individuals and communities to make informed choices about food production, consumption, and utilization, thereby contributing to healthier, more resilient, and more food-secure societies worldwide. As we strive towards this shared vision, let us remain committed to collaborative action, evidence-based decision-making, and inclusive approaches that leave no one behind in our pursuit of a world free from malnutrition.

Index